Race to the Finish?

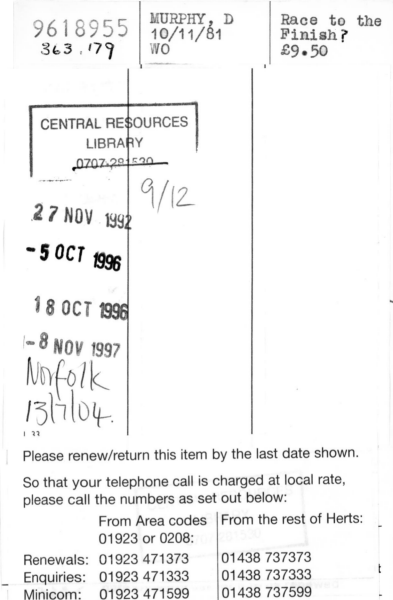

Please renew/return this item by the last date shown.

So that your telephone call is charged at local rate,
please call the numbers as set out below:

	From Area codes 01923 or 0208:	From the rest of Herts:
Renewals:	01923 471373	01438 737373
Enquiries:	01923 471333	01438 737333
Minicom:	01923 471599	01438 737599

L32b

Race to the Finish?

THE NUCLEAR STAKES

Dervla Murphy

JOHN MURRAY

For Brendan, who helped

© Dervla Murphy 1981

First published 1981
by John Murray (Publishers) Ltd
50 Albemarle Street, London W1X 4BD

Typeset by Inforum Ltd, Portsmouth
Printed in Great Britain by
The Pitman Press, Bath

British Library Cataloguing in Publication Data
Murphy, Dervla
Race to the finish?
1. Atomic Power
I. Title
363.1'79   TK9151.4
ISBN 0–7195–3884–X cased
ISBN 0-7195-3890-4 Pbk

Contents

Acknowledgements

Deborah Singmaster and Jeremy Bugler set me on the right research path when I didn't know which way to turn. Without their guidance, this book might never have happened.

Nor could it have happened without the help of those pro-nuclear experts who most generously and patiently gave of their time, knowledge and hospitality to an avowed anti-nuke — but who would probably prefer not to be named.

Without The Flowers Report ("Report on Nuclear Power and the Environment"), it would have been virtually impossible for an amateur like myself to embark on an organised study of the nuclear power controversy. It has been rightly remarked that Flowers can be quoted to equally telling effect by pro- and anti-nukes. Of all the books listed in the Bibliography, this is the volume I would most strongly recommend to any of my readers whose appetite for nukery may have been whetted by *Race To The Finish?*

From every continent came practical help and moral support. To the scores of people who contributed to this effort by sending books, newspaper cuttings, scientific journals, theses (published and unpublished), transcripts of lectures and court cases, accounts of personal experiences and (much-needed) letters of advice and encouragement — my heartfelt thanks and apologies for not having written separately to each one.

To those long-suffering friends on both sides with particular knowledge of specialist aspects of the controversy, who read, marked and returned various drafts of relevant chapters — I am especially grateful.

My editor, John R. Murray, expertly restored an exhausted author's mind before the final round of revisions and relentlessly deleted innumerable irrelevancies, inaccuracies and stupidities. (Those that remain are owing to authorial obstinacy.)

A suitable title eluded everyone — except Diana Murray, who put us all out of pain by coming up with a splendidly balanced title/sub-title.

At the eleventh hour, Jane Wingfield rescued both my printers and myself from almost certain nervous collapse by impeccably re-typing a mangled typescript during a sunny Easter week when she should have been enjoying herself.

My daughter Rachel uncomplainingly endured more than a year of physical neglect and emotional deprivation, while nukes took over the household. To her, my contrite gratitude — and I promise not to do it again.

Preface

Dr John Gofman, asked why he had become so involved in America's nuclear debate, said: "This nuclear thing, it was a stone that fell in my path, and before I could go on I had to kick it out of the way." Which is precisely how I have felt while writing this book, when I should have been writing about Peru.

The nuclear threat has already been defined by many competent experts but few of these are established authors. Therefore their books are not as well known as they deserve to be and the need to reach a wider readership inspired some of my friends to drive me into the arena. At first this seemed a nutty idea, given my peculiarly unscientific and untechnological make-up. However, it was argued that that apparent handicap might prove an advantage: anything scientific, written in language simple enough for *me*, must be generally understandable. Also, an anti-nuke book by a travel-writer might lure into the debate certain readers who had never previously concerned themselves with such issues.

My publisher, being a fair-minded man, longed for an "objective" book. In the 1980s, this is rather like looking for an objective book about White Slavery or the heroin trade. The nuclear weapons/power industry is capable of exterminating all life on earth: so one tends to write about it with a certain lack of detachment. However safely nuclear power stations may be designed, built and operated, there is no possibility of halting the spread of nuclear weapons while commercial nuclear technology is being vigorously marketed all over the world. Hence the majority have by now come off the fence and those who are not pro-nuke are "anti" — and proud of it.

The matter is urgent. This unique danger presents a unique challenge to human wisdom. A new technology has never before been abandoned and of course it goes against man's exploratory nature to seem to *retreat* by turning away from the Fissile Society just because novel difficulties have arisen. But the nature of those difficulties makes retreat essential. As Sir Kelvin Spencer (Chief Scientist to the Ministry of Power from 1954 to 1959) has put it: "Mankind is not grown-up enough to be trusted with nuclear power and the handling of plutonium waste — the most lethal, toxic, man-made element on earth."*

My first public appearance as an anti-nuke was in *Blackwood's Magazine* in September 1979. The resulting correspondence suggested that being an effective anti-nuke was not as simple as I had imagined. My article had been based on many newspaper cuttings and journals, and a few books — all acquired in the US. What I had not yet acquired was the ability to distinguish between reliable and unreliable sources: and the struggle to remedy that defect was long and hard. The Bibliography lists those books that have helped me most. And my reading was reinforced by visits to nuclear power stations and other nuclear installations, and by long conversations with a variety of eminent and kindly pro-nukes — chemists, physicists, engineers, radio-biologists and safety experts.

In February 1980 Sir John Hill, then Chairman of the United Kingdom Atomic Energy Authority, wrote an article for *Blackwood's Magazine* contesting my view of the nuclear power industry. The following passage, especially, encouraged me to persevere with the writing of this book.

"Nuclear power is a substantially cheaper way of producing electricity than the burning of coal. It is also cleaner, it is to be preferred environmentally, and it is safer. Why then are the developed countries of the world who have nuclear industries not getting ahead faster with their nuclear programmes? And why is there so much public concern about safety and accidents? One reason is articles like Dervla Murphy's. It is a common pattern of anti-nuclear articles to detail a long list of difficulties, delays and mishaps that have occurred in the world nuclear industry over the past twenty years, with the added question 'But what if . . .?' in relation to every one. This approach, if applied to other industries, would persuade us that the world is too dangerous a place to live in and that it is a pity life ever started on this planet."

Sir John's is a plausible argument, until one remembers that

*Daily Telegraph, 30 October 1980.

no other industry has the power to do such widespread and last-
ing damage should something go wrong. This is why anti-nuke
writers focus mercilessly on the "difficulties, delays and mishaps"
of the nuclear industry. Of course we realise that *every* human
enterprise must have its quota of mishaps; but the nuclear industry cannot afford even trivial accidents.

Despite the publicity given to the nuclear power controversy,
many people remain reluctant to consider it in detail. And they
are even more reluctant to think about nuclear war. Some years
ago, Professor Robert Jay Lifton suggested reasons for this: "In
Hiroshima, a defence mechanism called forth by people exposed
to the horrors of the bomb was simply to cease to feel, that is, to
become psychologically desensitised. Virtually every survivor I
spoke to . . . described being fully aware of what was going on
around them, knowing that people were dying in horrible ways,
but simply losing their emotional involvement in it all . . . This
form of psychic closing-off or psychic numbing was necessary, as
it is for anyone exposed to extreme catastrophe. The difficulty
was that the numbing process did not necessarily end when the
danger was over . . . Then I began to think of psychic numbing as
a much more general problem. For instance, a degree of psychic
numbing is necessary for anyone who deals professionally with
death. One can draw analogies with rescue workers of various
kinds, and with physicians in general . . . The paradox in all of
these situations lies in the way in which the same process can serve
as beneficial protector or harmful "deadening", depending upon
its appropriateness and degree. But I want to stress here the way
in which psychic numbing is called into play in relation to the
making, testing and anticipated use of nuclear weapons. Indeed,
so pervasive has psychic numbing become, much of it in relation
to these weapons, that rather than the age of anxiety we could well
speak of the Age of Numbing. To put the matter simply, one
cannot afford to imagine what really happens at the other end of
the weapon. I would stress the widespread, indeed universal
nuclear numbing affecting us today. It is more than a defence
mechanism. It really amounts to a re-orientation of the entire self,
with a muting of overall response to the nuclear environment.
It is another form of impairment to the symbolizing process,
especially to that aspect of it which ordinarily harmonises cogni-

tive and emotional elements — what one knows and what one feels. Thus we can now speak of knowledge without feeling."*

Recently several friends have expressed the fear that my writing an anti-nuke book will cause me to be associated in the public mind with "spurious ideologies", as one of them rather endearingly put it. This would indeed be regrettable, because as far as I'm aware I don't possess an ideology and never have — or not since my late teens when I was a fanatical Irish nationalist. (Is that an ideology?)

This lack should, I suppose, disturb me: obviously all the best people are ideologically equipped, one way or the other. But perhaps being an econut *is* having an ideology — by spontaneous combustion, as it were, rather than through any process of intellectual selection. Certainly, my apolitical career must make it difficult for anybody to accuse me of deliberate subversion. (A popular pro-nuke poisoned dart.) But I may well be labelled: "Unconscious Dupe of International Communism", or something of the sort. I'll survive — in good company. Bertrand Russell, I seem to remember, acquired a similar label during his CND phase. To econuts, all such labels are innocuous because meaningless in relation to the future. If we ever escape from the nuclear morass, our destiny is unlikely to be determined by either Capitalism or Communism as we now know them.

This book is very much the result of *other people's* research and study: while working on it I have often felt myself to be an editor rather than an author. It therefore seems right that the material gains should go to an appropriate good cause: The Irish Arts Council's Pension Fund for Artists. Whatever other gains there may be will go, I hope, to the whole of mankind.

---

*Robert Jay Lifton, *Boundaries*, Simon and Schuster, 1967.

1

In the Beginning was the Bomb

The freedom achieved by modern man is not unlimited. He is
not free to believe wholly what he wishes or do wholly what he
likes. He is free to make some changes in his world but he is not
free to create his world *de novo*. His world already exists. He
must come to terms with the world as it really is, and not as he, in
a flight of fancy, would prefer it to be.[1]

LLOYD GEERING

On 6 August 1945 I was thirteen and bored by war-talk. For
almost as long as I could remember people had been killing each
other in vast numbers all over the globe. Bullets and bombs — at a
safe distance — were part of my mental furniture. And Japan was
very far away.

Yet I do remember Hiroshima. I remember it as something
peculiarly dreadful, though not personally frightening. From my
parents' conversation I extracted an awareness that this was not
just another lethally brilliant military invention, like V2 rockets.
They seemed to regard the Atomic Age — headlined inches high
in every newspaper — as presenting an altogether novel philo-
sophical challenge to mankind. According to my father (something of a scientist *manqué*) this new era was *essentially* unlike all
those preceding tens of thousands of years of human history and
pre-history. His reasoning was beyond me, but the turbulent
atmosphere of that week has remained in my memory ever since.

There were conflicting emotional currents. On an official, political level ran the grisly exultation of the Allies, who had "Beaten
Germans in Battle of Science", as one headline declaimed — quite
forgetting that such names as Niels Bohr and Hans Bethe were
not particularly Anglo-Saxon. On another level there was quick
remorse. Eisenhower declared: "It wasn't necessary to hit them

with that awful thing." And of course there was fear. At a press conference in the White House, less than three hours after the bomb was dropped, President Truman announced: "British and American scientists, working together, have harnessed the basic power of the universe." But: "Further examination is necessary of possible methods of protecting us and the rest of the world from the danger of sudden destruction." It was explained that the implications of this harnessing, for both good and evil, were still hidden. (The thousands who, as the President spoke, lay dying in agony amidst the rubble of Hiroshima would scarcely have agreed.) The "swords into ploughshares" theme was also introduced by Mr Stimson, US Secretary of War, who referred to means that had been found to release atomic energy "not explosively but in regulated amounts". With those words, he had announced the birth of the nuclear power industry.

My father was not among those impressed by this harnessing of the basic power of the universe. The Greeks, he recalled, had intuited the atom. And Prometheus didn't do too well after stealing the fire of the gods— so maybe the Greeks had a message for us. But I was much too young to worry about the future of mankind. Soon I had dismissed the Atomic Age. To an unsophisticated adolescent, growing up in the happy depths of the Irish countryside, it seemed for quite a while supremely irrelevant. But during those years its relevance was rapidly increasing in the US, where the Manhattan Project, while producing the Hiroshima and Nagasaki bombs, had laid the foundations for both the nuclear arms race and the world's nuclear power programme.

The Manhattan Project, set up on 13 August 1942, has been described as "the greatest single achievement of organised human effort in history".[2] Under the almost frenetic leadership of General Leslie Groves, and at a cost of some $2 billion, it held its imaginary race with German scientists (who in fact were *not* making an atomic weapon) and tested its first atomic bomb on 16 July 1945.

During those three years, the military, academic and industrial establishments of America formed a close alliance which was, alas, to endure. At Oak Ridge, Tennessee, and around the then tiny village of Hanford in Washington State, elaborate laboratories and gigantic factories were built. Enriched uranium for the atom

bomb was produced at the Oak Ridge gaseous diffusion plant —
K-2 — which covered forty-four acres. K-2 had to be U-shaped
and each side of the building was half-a-mile long. It cost $500
million and none of the 200,000 construction workers knew of its
purpose. At Hanford, Du Pont produced plutonium: another
fissionable material for another sort of bomb. Radioactive contamination and high-level waste disposal were rarely mentioned;
little was known about such problems and only the Bomb
mattered . . .

Post-Bomb, many distinguished atomic pioneers wished to see
the atomic programme closed down. But America had been left
with some 600,000 highly trained atomic experts — industrial,
scientific, academic — and with an unprecedented complex
of vastly expensive plant that could be used only to produce
uranium, plutonium and atomic bombs. Splitting the atom had
become part of the American economy. So lots more bombs were
made and a commercial nuclear power programme was carefully
fostered by the Atomic Energy Commission (AEC). This is seen
by some as a way of justifying massive military expenditure.

The five-man civilian Commission, to be appointed by the
President every five years, was set up on 1 August 1946 and given
complete legal control of the uses of atomic energy, for both
military and civilian purposes. It was supervised by the Joint
(Congressional) Committee on Atomic Energy (JCAE), made up
of nine senators and nine representatives who were to be kept
informed of all AEC activities. This AEC/JCAE team has been
described as "one of the nation's major legislative disasters. In
their zeal to put atomic energy under civilian control, the legis-
lators merely totalitarianized nuclear power."[3]

In December 1953 Eisenhower outlined to the UN his government's "Atoms For Peace" programme. By then I was of an age to
be dogmatic about everything, including atomic energy, and this
programme did not impress me. It proposed the sale of enriched
uranium to other countries for the development of nuclear
power, and Eisenhower was confident that America's willingness
to share atomic know-how would work wonders for international
peace, prosperity and goodwill. An American political journalist
more realistically referred to the programme as "part of our Cold
War strategy. It was first advanced as a key to an increased

standard of living for underdeveloped areas. From the point of view of foreign policy the US was attempting to alter the image the world had of the US. We sought to divest ourselves of the stigma with which we had been tagged. Abetted by Soviet propaganda, the world had come to think of us as the nuclear bully preoccupied with the art of dropping megaton-size bombs."[4] From that day to this, the offer of nuke know-how — however ill-equipped the recipient may be to use it — has been one of America's favourite foreign policy tools.

Thirty-five years after Hiroshima, Mr Anders Thunborg, Sweden's Permanent Representative to the UN, observed: "The technological development of nuclear weapons systems is defended on the grounds of national security. But it is also believed that new weapons systems emerge not because of any military or security considerations but because technology by its own impetus often takes the lead over policy, creating weapons for which needs have to be invented and deployment theories have to be adjusted."[5]

The road along which mankind has been led since 1945 forces one to question Professor Michael Howard's argument: "My own view is that international conflict is an ineluctable product of diversity of interests, perceptions and cultures: that armed conflict is imminent in any international system; but that war can and must be averted by patience, empathy, prudence and the hard, tedious, detailed work of inconspicuous statesmanship — qualities which are notably absent from populist movements whose universal characteristic is a desire for the instant and total satisfaction of their demands. The appalling consequences of failure in the nuclear age make the exercise of these prudential qualities more vital than ever before. Romantic gestures will do nothing to help."[6]

Later on, we will consider the extent to which the nuclear industry (weapons plus bombs) has been allowed to foul up global development, thus setting the scene for international conflicts based on First World greed and Third World need. No "inconspicuous (or conspicuous) statesmanship", or "prudential qualities", have ever been effectively exercised to halt this process.

In contrast, the "romantic gestures" of various "populist movements" have produced some minor yet not insignificant results over the past decade. As Pierre Wack has pointed out: "The difficulties encountered by companies and state agencies in developing large-scale projects have seriously increased in the past ten years or so. Consider the history of just a few examples: the Alaska pipeline, the British Windscale reprocessing plant, nuclear plants in France and Germany, the UK Anglesea terminal and pipeline, the Anglo-French Concorde and the US supersonic transport. On the US East Coast, more than a dozen refinery construction projects have recently been blocked by local and/or state action based on environmental grounds ... In Holland, the process of securing a permit for the construction and operation of a new process plant required ten man-days of effort in 1965–6. In 1971–2, this had increased to two man-years; and by 1975–6, to four or five man-years per permit. In Germany, to obtain the *Genehmigung für Bau* takes two-and-a-half years in the most favourable case. If public complaints are received, the consequent processes can take up to five years ... People today are much less willing than in the past to accept decisions affecting them in which they have played no part. The phenomenon of participation — like decolonisation — is irreversible and is expected to develop further."[7]

To put it in an econutshell: *individuals* are on the move. Which is why it has seemed to me worth writing this book, to try to increase the numbers of individuals who are sufficiently aware of the nuclear threat, both military and civilian, to take action.

Strangely, many of our leaders are encouraging us to *fear* radical changes in our present way of life, instead of preparing us to accept them as inevitable — and educating us to recognise them as good. We, and they, have become cogs in a giant money-making machine whose operators well know how to keep everybody playing the Consumer Game. Thus the majority have been persuaded that nuclear power is an "advance" without which, as oil supplies dry up, we must revert to conditions of extreme hardship; so anti-nukes are reactionary and unrealistic.

In India, England, Peru, France, America, Ethiopia and Ireland I have come up against similar resistance to learning about nuke dangers. A common ploy is to make the anti-nuke feel foolish. "What! Don't tell me *you've* gone all freakish and cranky!

Be your age — you're behaving like a student in search of a cause! We've got to be practical. Remember what people said when trains were invented? That people would die because humans weren't meant to travel so fast! Get a grip on yourself — we have to live where we're at!"

All this is deflating, especially for anti-nukes who are neither eloquent nor temperamentally fanatical and who dislike forcing personal convictions on others, even for their own good. To my shame I often give up, in conversation, before I've really started — partly through sheer despair at the magnitude of the anti-nuke's task, partly through a reluctance to seem irrationally prejudiced.

It takes a whole book to explain why one cannot adopt a middle-of-the-road position on the nuclear controversy, as one would try to do on Northern Ireland, or South Africa, or the Arab-Israeli conflict. Indeed, that very affection for *people* which makes it easy to be objective about historical/political problems makes objectivity impossible when dealing with nuke technocrats. Not that they should be blamed, as individuals. They are merely acting according to their instincts, either as military men who believe in preparing for war, or as industrialists to whom nuclear fission is just another money-spinner. Clearly they all suffer from "nuclear numbing", as described by Professor R. J. Lifton (see my Preface). But the threat their conduct of affairs presents to mankind must inhibit us from being lenient towards them. As they are utterly unscrupulous in upholding nukedom, so we must be uncompromising in our opposition.

Most people (including many who are sympathetic to the anti-nuke cause) prefer not to dwell upon the nuclear future — and I can't say I blame them. Moreover, they may be wiser than the active anti-nukes; they may know in their hearts that a monster has been released which cannot be captured and rendered harmless. The anti-nukes, who so often sound hysterically pessimistic, are in fact the optimists. They still believe that it is worth while hunting the monster, however unequal the contest may look.

Often one merely skims newspapers or listens with half an ear to news broadcasts. When snippets about the more sensational nuke dangers are thus received, the non-human figures involved give the whole subject an unreal, abstract, even irrelevant, aura.

For instance: "Plutonium-239 continues to emit deadly radiation for some 240,000 years. Its iron-like properties enable it to cross the protective placental barriers and endanger a developing foetus. It can concentrate in the gonads and if so concentrated causes irreversible genetic mutations" — and so on. Such bizarre information is not easily assimilated; it fails to inspire a feeling that this is something about which individuals should exert themselves. People may think: "How awful!" But they also think: "So what!" Which suits the pro-nukes, who try to create a climate of opinion in which apprehension about low-level radiation seems derisory: not an emotion to which any well-balanced, responsible citizen would ever give way.

Yet many scientists worry about the possible genetic effects of low-level radiation accumulating throughout the environment — altogether apart from the hazards of high-level radiation, which nobody can deny. But here again we find "ostrichism" among the general public: nobody wants to think about mutated grandchildren or great-grandchildren. In everyday life we take sensible precautions to protect our young; this is a personal, private thing, an intimate duty which is also a joy. But most of us are not adapted to doing that duty more indirectly and publicly, on our descendants' behalf, by marching, protesting, wearing badges, staging sit-ins. Such exhibitionist campaigning seems remote from — almost alien to — the function of the average conscientious parent. The Atomic Age, however, has changed everything. On 6 August 1945 man was deprived of the most stabilising part of his heritage, his absolute certainty that "life goes on". After Hiroshima, the *New York Herald Tribune* described the atom bomb as "weird, incredible and somehow disturbing . . . One forgets the effect on Japan or on the course of the war, as one senses the foundations of one's own universe trembling."

To grasp the nature of the effects of applying nuclear technology takes some effort of the imagination. It is easier simply to accept nuclear fission at the pro-nukes' valuation — as part of the whole process of scientific discovery, of rising (for the West) standards of living, of exploitation of the earth's resources for the benefit of mankind, and so on. But nuclear fission is *not* just another marvellously convenient if abstruse discovery, like antibiotics, aeroplanes and television. It is *not* another mighty stride

towards the complete mastery of this planet. Rather is it a retro-
gression towards the time when man was at the mercy of overwhelming epidemics which he did not understand and could not
control. Although that is a weak analogy, for never before has our
race been exposed to indestructible, invisible, intangible, man-
made influences with the power to destroy or deform future
generations.

So it is that parents must now protect their children, and their
children's children, by going against the nuclear tide, by somehow forcing their governments, if they are fortunate enough to
live in a democracy, to halt the expansion of nuclear programmes
(weapons and power) and divert some of the money thus saved to
alternative energies.

Anti-nukes are frequently accused of being too emotional, as
though it were somehow improper to feel deeply about poverty
and disease, war and peace, life and death. Often pro-nukes seem
to forget that emotions are an integral part of the human make-
up and that only their *control* is a virtue. Their *suppression* is a
denial of our humanity, a psychic self-mutilation. Yet emotions
must be excluded from the technocrats' world. Otherwise they
could not continue to live (or refuse fully to live) as they do —
slaves to what Theodore Roszak has called the "myth of objective
consciousness".

Admittedly, anti-nukes, too, are an odd lot. The French sociol-
ogist, Alain Tourain, has shrewdly analysed the movement: "It is
only on the threshold of the new society, not inside it. Our social
and political life is still governed by the problems and forms of
action appropriate to an industrial society, rather than to a post-
industrial society, or one entering a new phase of industrial-
isation. It's hard for new themes of action to find political expression . . . It is because the anti-nuclear movement is ahead of its
time that it seems to be a cross between a social movement and a
cultural one; that it has a tendency towards Utopianism; and that
much of its activity goes into challenging ends, rather than looking for means . . . The construction of a new social movement is
difficult. It must dig its foundations in unstable soil, and it always
risks being reduced to a mere intellectual exercise or the fanatical
faith of a sect . . . Ecologists are very far from being an organised

political force, but they attract a good number of people who no longer feel represented by the parties. Half-way between protest and politics, they are the explorers of an unknown country, who have returned from their expedition dazzled and exhausted by what they have discovered and experienced . . . They are lost in the unseeing crowd, because they are the only ones who look into the future."[8]

The abandonment of environmentally detrimental High Technology is as yet inconceivable to most people; and obviously the colossal, intricate machine of a modern industrial society cannot be dismantled overnight. We must withdraw slowly from this trap into which we have been lured by our own cleverness, and Professor Leopold Kohr has suggested how this might be done. "Technological *humanists* try to avoid the toxic side-effects of High Technology by concentrating on the search for new means of production that won't use so much energy. A simpler, fuel-saving, intermediate technology means, because of its lower mechanical efficiency, that the highest standards of living can be achieved *only* if all hands are employed. In other words, inter-mediate technology will solve not only the energy problem, but also the socially infinitely more threatening unemployment problem of automated technology . . ."[9]

Anti-nuclearism is often part of a general hostility to High Technology. The Flowers Report (para. 499) noted: "It seems that nuclear power has in some ways become the whipping-boy for technological development as a whole. Many thoughtful people believe there is a progressive deterioration of the environ-ment which they associate with the spread of technology . . . but they feel unable to make a stand against this process because each individual step is small and can probably be justified as bringing benefits at minimal environmental costs. Nuclear power provides a dramatic focus for opposition in some countries to technological development, and we have no doubt that some who attack it are primarily motivated by antipathy to the basic nature of industrial society and see in nuclear power an opportunity to attack that society where it seems likely to be most vulnerable, in energy supply."[10]

This is a fair analysis of the motives of many anti-nukes. Yet it should be emphasised that the several *unique* disadvantages of nuclear power are making this form of electricity generation unacceptable to increasing numbers who have no particular grudge against industrial society — and may indeed be pillars of it.

The ramifications of the nuclear controversy are formidable. It is no mere difference of opinion about how to generate electricity, but a future-shaping debate on the path mankind should take at the most crucial cross-roads in human history. Are we to hurry on towards an intolerably tension-filled, centralised, unemployed, polluted and unjust world, in which the spread of nuclear power makes ever easier the proliferation of nuclear weapons? Or should we acknowledge the growth/affluent/consumer society as an experiment that has failed, and turn aside into the unknown, to strive gradually to reform our civilisation in a manner showing reverence for all of creation?

If we do persist on our present path, until disaster forces us off it, our future crimes will be the more heinous because deliberate. While some scientists have been despoiling the earth, others have been exposing this abuse. Now we know what we are doing, as our forefathers did not when the Industrial Revolution began — an uncontrolled explosion of Western man's inventiveness, enterprise, vigour and greed. Within the past generation people's expectations have been dramatically altered to suit the proponents of the growth society; at some stage they must be altered again, to suit the planet we live on. As the editors of *The Ecologist* have pointed out: "It should go without saying that the world cannot accommodate this continued increase in ecological demand. *Indefinite* growth of whatever kind cannot be sustained by *finite* resources. This is the nub of the environmental predicament."[11]

2

Murphy's Law at Three Mile Island*

If anything can go wrong, it will go wrong.

MURPHY'S LAW

It was 2.30 p.m. on Tuesday, 27 March 1979. Within a few miles of my Greyhound bus, as it sped along the Pennsylvania Turnpike, the free world's most serious commercial nuclear reactor accident was about to happen. My companion noticed the signpost pointing to Three Mile Island (TMI) but I missed it. We had left Chicago at 8 p.m. the previous evening and the undulating Pennsylvanian countryside was having a soporific effect.

Beside me sat a retired high-school teacher from Ohio with the curiously unwrinkled, lard-coloured, rectangular face of a certain type of elderly American male. Earlier, he had glowered at my anti-nuke badge— "Better Active Today Than Radioactive Tomorrow!" Plainly he deprecated such symbols of shallow romanticism about the environment, or paranoid nervousness about health, or lack of sympathy for the energy needs of an advanced society. He said, rather truculently, "So no nukes — then what do we use for power? Anyway coal's worse — a *lot* worse. Gives people bronchial troubles and kills miners. And if we release much more carbon dioxide into the atmosphere we'll destroy the global climate. But anti-nuke freaks can't think straight."

*Unless otherwise indicated, all quotations and figures given in this chapter are taken either from "The Report of the President's Commission on the Accident at Three Mile Island" (The Kemeny Report), obtainable from the Superintendent of Documents, US Government Printing Office, Washington DC 20402, *or*: from my personal notes of local wireless and television broadcasts, and newspaper reports, during the TMI crisis.

Nobody likes to be classified as a freak, least of all those with tendencies that way. I replied crisply that an awareness of coal hazards partly explains the anti-nukes' support for alternative energy research, which could by now have contributed significantly to our needs had it received even a fraction of the billions of dollars invested in nuclear energy over the past thirty years. Afterwards I wondered if the events of the next week did anything to modify my acquaintance's fissile enthusiasm.

The following morning, soon after six o'clock, I was sitting out in the sun drinking coffee and enjoying the freshness of my Pennsylvanian friends' garden. Meanwhile, some sixty miles away, it was being discovered that radiation levels were rising rapidly in the TMI-2 reactor. Things had been going seriously wrong there since thirty-six seconds past four o'clock, yet the staff were only then beginning to think that perhaps the outside world should be put on the alert. At 7.05 a.m., as I was walking across the silent dewy fields at William's Corner, Kevin Molloy, the Dauphin County Civil Defense director, was warned that an on-site emergency had been declared — as laid down by TMI's emergency plan, should events threaten "an uncontrolled release of radioactivity to the immediate environment". Molloy, driving towards his command post, next heard, on his mobile radio, that at 7.24 a.m. the plant had declared a general emergency in response to an "incident (*sic*) which has the potential for serious radiological consequences to the health and safety of the general public".

When things began to go wrong there were four men in charge of TMI-2: William Zewe, the shift supervisor in charge of both TMI-1 and TMI-2, Frederick Scheimann, the shift foreman for TMI-2, and the two control-room operators, Edward Frederick and Craig Faust, aged twenty-nine and thirty-two. Frederick had already dealt with one emergency shutdown at TMI-2 and he believed that its designers had prepared for every conceivable crisis; nearly half a billion dollars had been spent on the plant's multiple fail-safe protective systems and it was then popularly regarded as a superb example of the world's most sophisticated engineering technology. It had been operating for only three months, since 30 December 1978. Altogether almost a billion dollars had been invested in its design and construction and

Babcock and Wilcox (B & W) had devoted a decade to building it. Yet it had suffered serious growing-pains. In mid-January, two safety valves ruptured during a turbine test and it had to be shut down for two weeks. On 1 February, a throttle valve developed a leak. The day after, a heater pump blew a seal. On 6 February, a feedwater line pump tripped off; and the reason for its stutter was allowed to remain a mystery.

One has to sympathise with Frederick and Faust as the victims of an inadequate training system, a badly designed control room, an unreliable safety agency, an incompetent management and an unforgiving technology. Within moments of the first alarm sounding, *one hundred* other signals — the majority irrelevant — assailed these unfortunate men. Faust later told the President's Commission: "I would have liked to have thrown away the alarm panel. It wasn't giving us any useful information."

Fourteen seconds into the accident, one of the operators noticed that the emergency feed pumps were running; but he failed to notice two lights indicating a closed valve on each of the two emergency feed lines — which meant the steam generator's water supply was cut off. One light was covered by a yellow maintenance tag; presumably the second went unnoticed through panic. At that stage Zewe, who had been working in a tiny, glass-enclosed office behind the operators, alerted the TMI-1 control room and summoned the TMI-2 shift foreman.

Scheimann had been supervising a maintenance job on the plant's no. 7 polisher, a machine to remove dissolved minerals from the feedwater system. His workmen were using a mixture of air and water to break up resin that had clogged a resin transfer line. Later it was discovered that a faulty valve in one of the polishers allowed water into the air-controlled system that opens and closes the polishers' valves — which may have caused their sudden closure just before 4 a.m., which in turn may have caused the pump stoppage that initiated the accident. On at least two previous occasions water had leaked into the polishers' valve control system but the plant's owners, Metropolitan Edison Co. (Met Ed), had not corrected this fault.

When something goes wrong in a British Magnox or AGR reactor their much more highly qualified operators have time to think; when something goes wrong in an American LWR, the time-scale makes inhuman demands. Thirteen seconds after the TMI accident began, the operators turned on a pump to add water to the system; thirty-five seconds later, the water in the pressuriser began to rise again; fifty seconds later, because of the blocked emergency water lines, the steam generators boiled dry; fifteen seconds later, pressure in the cooling system dropped sharply and two high-pressure injection emergency pumps began to pour some 1,000 gallons a minute into the system. Two and a half minutes later Frederick misinterpreted the signals, turned off one pump and reduced the flow of the other to less than 100 gallons a minute. There were, you will recall, a prodigious number of signals demanding virtually simultaneous attention.

At about 4.20 a.m., a telephone call from the auxiliary building warned Frederick that an instrument there indicated more than six feet of water in the containment-building sump. He consulted the control-room computer, which confirmed this. As he had no idea where the water was coming from, and didn't want to endanger the public by releasing possibly radioactive water, he advised shutting off the two sump pumps in the containment building. Both were stopped at about 4.39 a.m., by which time 8,000 gallons of radioactive water had been pumped into the auxiliary building.

For over four hours the operators continued to misjudge the level of water required to cope with the accident and remained unaware that their actions could result in an uncovered core — and then a meltdown. It later emerged that by 6.48 a.m. two-thirds of the twelve-foot-high core stood uncovered. At this time, in certain parts of the core, the temperature rose to between 1900 C° and 2200 C°. Meltdown is thought to occur at about 2900 C° (about twice the temperature required to melt ordinary steel). Nobody knows for sure; one doesn't experiment with meltdowns.

At the meltdown temperature the material in the core would melt its way through the reactor floor and the earth, penetrating rock and any other substance until, on striking water, it exploded upwards like a volcano, spreading radioactive debris for miles. This is similar to what happens when lava hits water inside a volcano. But unlike molten rock the material inside the core is continually *generating* heat, so its penetrating ability is vastly greater than molten rock at the same starting temperature. Thus the explosion would be greater and radioactive material would be spread over a vast area.

Reading the Kemeny Report gave me my first nightmare in twenty years. I woke sweating one dawn, having dreamed that my kitchen had become a malfunctioning nuclear-power-station control room. Consider, for instance, the following extract:

"The two twelve-valves were known to have been closed two days earlier, on March 16, as part of a routine test of the emergency feedwater pumps. A Commission investigation has not identified a specific reason as to why the valves were closed eight minutes into the accident. The most likely explanations are: the valves were never reopened after the March 26 test; or the valves were reopened and the control-room operators mistakenly closed them during the very first part of the accident; or the valves were closed mistakenly from control points outside the control room after the test. The loss of emergency feedwater for eight minutes had no significant effect on the outcome of the accident. But it did add to the confusion that distracted the operators as they sought to understand the cause of their primary problem."

I should think so, too.

In its *Overview*, the Kemeny Report noted: ". . . while equipment failures initiated the event, the fundamental cause of the accident was 'operator error'. If the operators (or those who supervised them) had kept the emergency cooling system on through the early stages of the accident, TMI would have been limited to a relatively insignificant incident . . . Let us consider some of the factors that significantly contributed to operator confusion. First of all, it is our conclusion that the training of TMI operators was greatly deficient. Second, we found that the specific operating procedures, which were applicable to this accident, are at least very confusing and could be read in such a way as to lead the operators to take the incorrect actions they did. Third, the lessons from previous accidents did not result in new, clear instructions being passed on to the operators . . . The legal responsibility for training operators and supervisors for safe operation of nuclear power plants rests with the utility. However, Met Ed, the GPU subsidiary which operates TMI, did not have sufficient expertise to carry out this training programme without outside help. They, therefore, contracted with Babcock & Wilcox, supplier of the nuclear steam system, for various portions of this training programme. While B & W has substantial expertise, they had no responsibility for the quality of the total training programme, only for carrying out the contracted portion. And coordination between the training of the two companies was extremely loose."

At 4.45 a.m. George Kunder, superintendent of technical support at TMI-2, arrived at the Island, having been informed of the turbine trip and reactor scram. He was baffled by what he found and later told the President's Commission: "I felt we were experiencing a very unusual situation, because I had never seen pressuriser level go high and peg in the high range and at the same time, pressure being low. They have always performed consistently." The control-room staff supported this view: they described the accident as a combination of events they had never experienced, either in operating the plant or in their training simulations.

The four reactor coolant pumps began to vibrate violently soon after five o'clock but no one recognised this as another sign that the water was boiling into steam. There was a general fear that the severity of the vibrations might damage the pumps or coolant piping, so two of the pumps were shut down at 5.14 a.m. and the others at 5.41 a.m. By six o'clock it was clear that some of the fuel rod claddings had been ruptured by the high gas pressures inside them and that the coolant water had been contaminated by escaping radioactive gases. Because coolant was continuing to pour from the open pilot-operated relief valve (PORV), and little water was being added, the core top was exposed and became so overheated — as we have already seen — that the zirconium alloy of the fuel rod cladding reacted with steam to produce hydrogen. Some hydrogen escaped into the containment building; some — of which we'll be hearing a lot more — stayed within the reactor.

Soon after six o'clock, George Kunder organised a telephone conference with John Herbein, Met Ed's vice-president for generation; Gary Miller, TMI station manager and Met Ed's senior executive stationed at the nuclear facility; and Leland Rogers, the B & W site representative at TMI. Giving evidence before the President's Commission, Rogers explained that in the course of this discussion he asked if the block valve between the pressuriser and the PORV, a backup valve that could be closed if the PORV stuck open, had been shut. Kunder replied, "I don't know," but at once sent someone to find out. Only then was the block valve shut: at 6.22 a.m., two hours and twenty-two minutes after the PORV had opened, causing thousands of gallons of highly radioactive water to flood the floor of the pump house — an unsealed building from which several million curies of radioactive gas poured through ventilators soon after midday. At this point one wonders why the operators did not SOS for high-powered outside help the moment the accident started — or rather, why there is no regulation insisting that they should do so, however trivial the initial difficulty may seem. It is a disquieting thought that TMI-2, on 28 March 1979, had no one on the premises, or anywhere near, who was sufficiently qualified to deal with what began as a minor problem. How many other American reactors are similarly mis-staffed?

It has not yet been definitely established whether Leland Rogers, or someone else, should get the credit for the crucial valve being at last closed. Frederick soon after testified that a shift supervisor coming on the next shift suggested checking it; but at a Commission hearing on 30 May 1979 he declared that it was closed because the operators could think of no other way to regain control of the reactor — which suggests that they had forgotten its existence and purpose until 6.22 a.m. Whoever was responsible for closing it, the loss of coolant was now stopped and pressure began to rise; but so much damage had already been done that the crisis worsened. And it was not eased by the fact that for some utterly inexplicable reason almost an hour passed before the high-pressure injection was used to replace the lost water. By that time Kunder and his colleagues had somewhat belatedly realised that they were on the edge of disaster. When the operators turned on one of the reactor coolant pumps at 6.54 a.m. severe vibrations again caused it to be shut down, at 7.13 a.m. And for an hour monitors had been showing rapidly rising levels of radiation in the containment and auxiliary buildings, while radiation alarms had been going off in every direction.

Gary Miller, the TMI station manager, who had been told of the trip and scram within moments of their occurring, arrived on the scene soon after seven o'clock. Finding that a site emergency had been declared, he took over as emergency director and at 7.24 a.m. declared a general emergency. By then the containment-building radiation level had risen to about 800 rems *per hour*. (The maximum permissible radiation dose for the British public is 0.5 rems *per year*.)

Other TMI officials were now crowding into the control room, including Richard Dubiel, supervisor of radiation protection and chemistry; Joseph Logan, superintendent of TMI-2; and Michael Ross, supervisor of operations for TMI-1. But nobody had any constructive advice to offer. By 7.30 a.m. there were some forty technicians milling about: too many, perhaps.

Meanwhile, various State authorities had been alerted. But it proved difficult to notify the Nuclear Regulatory Commission's Region I office in King of Prussia — just a few miles from where I was at that very hour happily bird-watching. An answering service took the first telephone call and neither the NRC duty officer nor the regional deputy director could be contacted at their homes; both were driving to work. It was 7.45 a.m. before the NRC learned of that accident which was to lead the President's Commission to such an unflattering conclusion: "There is no well-thought-out, integrated system for the assurance of nuclear safety within the current NRC."

Half-an-hour earlier, emergency workers had had to evacuate the auxiliary building at TMI-2. William Dornsife, a nuclear engineer with the Pennsylvania Bureau of Radiation Protection, was then on the telephone to the control room. He heard the evacuation order and later recalled: "I said to myself, 'This is the biggie!'"

And so it could have been, though Met Ed remained very reluctant to admit that fact. By 11 a.m. radiation levels in the control room were "unacceptable"; everyone had to use protective face masks, with filters to catch airborne radioactive particles. These made communication difficult and may well have contributed to what the Kemeny Report — using more circumspect language — diagnosed as a ball-up.

All morning, operators and supervisors by the score struggled to subdue their antagonist. When every effort had failed they began, at 11.38 a.m., to decrease pressure in the reactor system, which caused another loss of coolant and uncovering of the core. Three workers then entered the auxiliary building and measured radiation levels from 50 millirems to 1,000 rems (i.e., one million millirems) per hour. Each received an 800 millirem dose.

At 12.45 p.m. Route 441, near TMI, was closed to traffic by the State police at the request of the Bureau of Radiation Protection.

Meanwhile, soon after 11 a.m., I had heard on the local radio station the first of those ambivalent, contradictory or downright inaccurate "expert" statements that during the next week were to unnerve Pennsylvania and disturb the whole world. Lieutenant Governor William Scranton III, the recently elected chairman of the Pennsylvania Emergency Management Agency, had called a press conference to inform the public that he had been assured by Met Ed's utility vice-president, Jack Herbein, that the problem was "well in hand". In view of the chaos then prevailing in TMI-2's control room, one can only conclude that someone had decided to put loyalty to Met Ed— and the whole nuclear industry — before the public safety. An unnecessary mass evacuation of the area, involving some 900,000 people, would have had disastrous long-term consequences for the industry. (The Kemeny Report noted: "We do not find that there was a *systematic* attempt at a 'cover-up' by the sources of information." My italics . . .)

Shortly after noon, as the first burst of radioactive gas rose above TMI, another Met Ed official, Don Curry, was informing the world: "The plant is cooling down in an orderly manner, with no consequences to the public." It all depends on what you mean by "orderly" . . . Less than two hours later, at 1.50 p.m., everyone in the control room heard a strange noise, later described by George Gary Miller as "a thud". That was the hydrogen exploding inside the containment building. A computer strip chart recorded the explosion's force: twenty-eight pounds per square inch. Within moments Met Ed's Michael Ross had examined the chart, yet nobody recognised the significance of the event. B & W's Leland Rogers told the President's Commission that at the time "the noise was dismissed as the slamming of a ventilation damper". As for the chart, Ross explained to the Commission: "We kind of wrote it off as possibly instrument malfunction."

Other instruments were in fact malfunctioning by this time, as the Kemeny Report records. "Several instruments went off-scale during the course of the accident, depriving the operators of highly significant diagnostic information. These instruments were not designed to follow the course of an accident. The computer printer, registering alarms, was running more than two and a half hours behind the events and at one point jammed, thereby losing valuable information."

At 4 p.m., William Scranton called another press conference, prompted by Governor Dick Thornburgh, a lawyer with an engineering degree. Thornburgh longed to be able to believe the Met Ed officials but as an experienced State Prosecutor could not evade the fact that the public was being taken for a ride. Scranton's statement on this occasion warned the press: "The situation is more complex than the company first led us to believe ... Met Ed has given you and us conflicting information." This statement was based on Scranton's meeting with government inspectors from the NRC, who had revealed that the fuel rods were damaged.

During the forenoon the NRC had begun to monitor radiation levels around the plant; to the south they took a reading of 30 millirems per hour and then added to the confusion by declaring that this was "not an indication of a serious problem", though the average American is exposed to only about 100 millirems per *year*.

Human nature is inscrutable — even one's own share of it. I am perhaps more aware than most of radiation hazards, yet I did not that day forego my afternoon walk among the local fields and woods, which were less peaceful than usual, with monitoring helicopters from the NRC Incident Response Centre at King of Prussia hedge-hopping all around me. Why does one stick stubbornly to one's normal pattern of life, when prudence suggests that it might be more rational to alter it? Do we feel a need to defy those powers (nuclear and otherwise) which are now eroding our personal liberties, as the sea nibbles at a coast-line?

Probably I was right to play it cool; and probably Jack Herbein was right to evade preparations for a general evacuation. According to the Kemeny Report: "In spite of serious damage to the plant, most of the radiation was confined and the actual release will have a negligible effect on the physical health of individuals. The major health effect of the accident was found to be mental stress." Yet one wonders about that word "negligible". The effects of low-level radiation on the human body are so little understood that in France the officially tolerated level is one hundred times greater than in the US. Dr Karl Morgan, for twenty years Chairman of the Internal Dose Committee of the International Commission on Radiological Protection (ICRP), has said, "We have made serious mistakes which we must correct." And George Wald, the Nobel Prize biochemist and Professor Emeritus at Harvard, believes that "every dose of radiation is an overdose".[1] It is to the nuclear industry's advantage that the modern world contains so many carcinogenic elements; by the time the long-term effects of any nuclear accident have become apparent, it may be impossible to attribute them positively to man-made radiation. But this subject needs a chapter to itself.

By that evening, "media people" from all over the world had begun to converge on Harrisburg. But I felt no urge to approach any closer to TMI than Fate had placed me — an idiosyncracy which has since greatly puzzled my media friends. To me, an out-of-control reactor looks like a Risk rather than a Scoop. If dramas were imminent, I reckoned William's Corner would be a much more salubrious vantage point than TMI.

The following day, 29 March, NRC officials informed us that the radioactive cloud around TMI was "no worse than one chest X-ray every six or eight hours". Some William's Corner residents reacted to this calming calculation by observing that one doesn't normally have chest X-rays three or four times a day — and anyway what about one's genitals? During that Thursday there were wildly disparate estimates of the amount of radiation that actually was escaping. The public might have found these less upsetting had someone explained that a cloud of gas, like any other cloud, is mobile and elusive and of ever-varying density. By that stage Met Ed's officials were — as Mike Gray afterwards wrote — "exuding all the confidence of fish sitting in a tree".[2] They had had to admit to fuel damage but were insisting that it was limited to 1 per cent. The NRC inspectors, however, advised their headquarters that some 18,000 fuel rods, almost 50 per cent, had probably been affected. And the Kemeny Report estimated the damage to be by then 90 per cent. It noted: "Although the NRC personnel were on-site within hours of the declaration of a site emergency and were in constant contact with the utility, the NRC was not able to determine and to understand the true seriousness and nature of the accident for about two days."

During that Thursday afternoon, Gordon MacLeod, Pennsylvania's Secretary of Health, and Anthony Robbins, Director of the National Institute for Occupational Safety and Health, had a telephone conversation that was to become a source of bitter recrimination and public disquiet. According to MacLeod, Robbins urged him to recommend an evacuation of people living around TMI. Robbins later denied that he had made any such suggestions, or even discussed evacuation. After this conversation, MacLeod arranged a telephone conference with the director of the Pennsylvania Emergency Management Agency, the director of the Bureau of Radiation Protection and one of Scranton's aides. He told them that Robbins had strongly recommended evacuation but they rejected the idea. MacLeod then argued for an evacuation of pregnant women and children under two, but this suggestion was also rejected. Yet at 2.10 p.m. a monitoring helicopter over TMI had detected a brief surge of radiation measuring 3,000 millirems per hour fifteen feet above the plant's vent. At NRC headquarters news of this reading was calmly received.

Another Thursday quarrel concerned the discharge of radioactive waste into the Susquehanna River. Early on the previous morning Met Ed had very properly stopped the routine discharge of waste water from lavatories, showers, laundries and leakages in the turbine, control and service buildings. Radioactive gases had contaminated it, though its radiation levels were still within the limits set by the NRC. By Thursday afternoon, however, the tanks contained 400,000 gallons and were almost overflowing. Two NRC officials then agreed to a release but, though Met Ed notified the Bureau of Radiation Protection, no communities living downstream from TMI were informed, nor was the press. When NRC Chairman Hendrie heard of the release he at once halted it — after some 40,000 gallons had reached the river. This was a tough nut to crack; everybody realised that an authorised release would infuriate the already jittery and understandably mistrustful public, yet if not emptied the tanks would simply overflow. The problem was debated for hours and on the late news I heard a press statement from the Department of Environmental Resources saying it had "reluctantly agreed the action must be taken". Soon after midnight, the release was resumed.

This incident well illustrates the *unique* hazards associated with nuclear power. However radioactive that water might have been, it had to escape to the environment — either through being released, or through an overflow. It was not controllable, for all the talk we hear about adequate protection for the public in the event of accidents.

While various Agencies, Bureaux, Commissions and Departments — ABCD, if I may be allowed my personal acronym — agonised over the Susquehanna, unpleasant things were being revealed back at the shack. During Governor Thornburgh's late afternoon press conference, Charles Gallina of the NRC had announced that all danger was over for people off the Island. The Governor sat looking thoroughly unconvinced and soon his scepticism was justified. At 6.30 p.m. Gallina and James Higgins, an NRC reactor inspector, were told the results of an analysis of TMI-2's coolant water. This confirmed core damage far in excess of anything the NRC had been prepared for. Higgins rang the Governor's office at 10 p.m. to admit that sporadic involuntary radiation releases were possible during the next few days. Within the reactor nothing had changed since the late afternoon; but within the NRC the dime had at last dropped.

It is unnerving to realise that the NRC "has primary responsibility and regulatory authority for health and safety measures as they relate to the operation of commercial nuclear plants". The Kemeny Report relentlessly exposed NRC ineptitude, ignorance and sloth. It is a bloated bureaucracy so enmeshed in its own red-tape that in no circumstances can it make prompt effective responses — even when reliable individual staff members point to some grave defect requiring immediate attention. It consistently plays the nuclear industry's game, though its theoretical function is to check on industry corner-cutting. The Kemeny Report was forceful: "With its present organisation, staff, and attitudes, the NRC is unable to fulfil its responsibility for providing an acceptable level of safety for nuclear power plants." Yet this was the organisation on whose word we had to rely during radiation releases from TMI-2.

Fortunately for the public peace of mind (or what remained of it), no one was then aware of the fundamental flaws in the NRC's evacuation strategy. Nor did they know that Met Ed had neglected to correct defects in the TMI monitoring equipment, though these had been pointed out by the NRC months previously. Equally — or more — serious, the TMI-2 iodine filters had been used continuously, instead of being reserved to deal with contamination, so these failed to work properly when the accident happened. Met Ed took maintenance lightly and the NRC was indulgent. After the accident, boron stalactites were found hanging from the valves in the TMI-1 containment-room building and stalagmites had built up from the floor. This might seem improbable, until you know that shortly before the accident an NRC inspector completed an inspection of TMI-2 *without looking at any of the equipment.* He merely talked to the staff and examined the records.

We remarked that evening how much more haggard and tense Governor Thornburgh looked each time he appeared on television, which he seemed to do every few hours. But his dark suit was always neat and relieved by a tiny, precise triangle of white handkerchief in the breast pocket. His receding hair was brushed back from a high forehead and his eyes were tired behind heavy spectacles. As the man ultimately responsible for local health and safety, he was being treated outrageously by Met Ed. All the evidence indicates that their on-site technicians, though possibly not their management personnel, were aware of the core exposure soon after it happened. Yet this vital information was withheld from the man whose duty it was to organise an evacuation should one suddenly prove necessary.

Some optimists imagined that by 30 March the NRC, Met Ed *et al.* would have got their act together. But not so. On that Friday confusion reached its climax, compounded by misinformation, fuzzy communications, increasing distrust between all concerned and a perverse coincidence. Only B&W kept out of the mêlée; they had resolved not to comment on the accident, even when their officials were aware of the media's being misinformed by others.

James Floyd was TMI-2's supervisor of operations for the midnight-to-noon shift: surely a long time to be in charge of a rebellious reactor. At 7 a.m. he decided, without consulting anyone else, that it had become essential to transfer radioactive gases from one tank to another — though this must release some radiation. (His reasoning may be studied in detail on page 136 of the Kemeny Report.) At 7.44 a.m. he asked for a helicopter to measure the escaping radiation. At 7.56 a.m. 1,000 millirems per hour were reported and at 8.01 a.m. 1,200 millirems per hour, 130 feet above the vent.

Then came the coincidence. "Shortly before 9 a.m., Lake Barrett, at the NRC Bethesda headquarters, was told of a report from TMI that the waste gas decay tanks had filled. He was asked what the release rate would mean in terms of an off-site dose. He did a quick calculation and came up with a figure of 1,200 millirems per hour at ground level. Almost at that moment, someone in the room reported a reading of 1,200 millirems per hour, just detected at TMI. By coincidence, the reading from TMI was identical to the number calculated by Barrett. The result was instant concern among the NRC officials. Communications between the NRC headquarters and TMI had been less than satisfactory from the beginning. Now NRC officials proceeded without confirming the reading and without knowing whether the 1,200 millirem per hour reading was on- or off-site, whether it was taken from a helicopter or at ground level, or what its source was. They would later learn that the radiation released did not come from the waste gas decay tanks. The report that these tanks had filled was in error."

After some discussion — time that would have been better spent determining the real state of play at TMI — Harold Denton, the NRC Director of Nuclear Reactor Regulation, decided to notify the Pennsylvania authorities that senior NRC officials recommended an evacuation order from the Governor.

Then ABCD really came into its own; from the public's point of view, the last vestiges of stability were replaced by a whirlpool of advice, counter-advice, argument, persuasion, denial, affirmation, recommendation and command. Harold Collins, Assistant Director for Emergency Preparedness in the Office of State Programmes, rang Oran Henderson, Director of the

Pennsylvania Emergency Management Agency, and advised an evacuation. Henderson rang Lieutenant Governor Scranton, who promised to ring the Governor. Thomas Gerusky, Director of the Bureau of Radiation Protection, denied that any evacuation was necessary; or any more necessary than it had been the day before. (He had had the wit to ring TMI.) Kevin Molloy, Director of Dauphin County Emergency Preparedness Agency, was told by Met Ed that no evacuation was necessary. But by then no one believed a word Met Ed said and when Henderson rang Molloy half-an-hour later, and told him to expect an evacuation order in five minutes, Molloy at once made a radio announcement, alerting the public. Shortly after, at TMI, NRC's Charles Gallina was disconcerted to meet an infuriated Met Ed employee who said, "What the hell are you fellows doing? My wife just heard the NRC recommended evacuation!" Gallina hastened away to check radiation readings on- and off-site, then rang Bethesda to try to cancel the evacuation notice. Governor Thornburgh almost simultaneously rang the NRC Chairman, Joseph Hendrie, who said no evacuation was needed but advised the Governor to urge everyone five miles downwind of TMI to stay indoors for the next half-an-hour. Thornburgh instead advised everyone within a ten-mile radius of the plant to stay indoors until further notice. And he begged for a *single expert* — just one, and a *real* expert — to be sent to TMI to pass on to him technical information and advice. An hour later President Carter, having talked with Hendrie, rang Thornburgh and promised to send an expert — Harold Denton, who had initiated the morning's chaos by recommending an evacuation without checking on the situation at TMI. The President also promised a hot-line between the White House, the NRC headquarters, the Governor's office and TMI.

Thornburgh then summoned his advisers to discuss the situation and one of Macleod's aides renewed the plea for an evacuation of pregnant women and small children. As he spoke Hendrie rang again, to apologise for the NRC error in recommending an evacuation, and Thornburgh repeated MacLeod's advice to him. Hendrie replied: "If my wife were pregnant and I had small children in the area I'd get them out because we don't know what is going to happen." Thornburgh put down the telephone and made up his mind. Soon after 12.30 p.m. I heard his message on the radio: "We recommend all pregnant women and pre-school children to leave the area within a five-mile radius of TMI and that all schools within that area should be closed."

At 11 a.m. Jack Herbein of Met Ed had held a press conference. The reporters knew exactly how much radiation had been released that morning but Herbein, apparently, didn't. He stated that the release had been measured at between 300 and 350 millirems per hour and, when challenged, protested, "I hadn't heard the number 1,200." From the start the press had suspected Met Ed officials of lying and now they were merciless in their questioning. Finally Herbein snapped: "I don't know why we need to tell you each and every thing that we do, specifically." One can sympathise with him there, since few journalists are capable of accurately reporting the details of a nuclear drama — particularly those relating to radiation. However, that remark demolished the last shred of Met Ed's credibility with the press.

A few hours later we first heard, from an NRC spokesman, the word "meltdown". At last Met Ed had been forced to admit that the core was *not* cooling, as they had so often claimed, and that there was "a remote risk" of a meltdown, described in nuclear jargon as a "class nine accident". Meltdowns are assumed to be *such* a remote risk that plant designers never even consider them. In March 1957 the Atomic Energy Commission published a document entitled "Theoretical Possibilities and Consequences of Major Accidents in Large Nuclear Power Plants" (WASH-740). Its 1964 revision estimated that a meltdown might "render an area equal to that of the state of Pennsylvania permanently uninhabitable" — a ghoulish coincidence which delighted journalists during the TMI crisis. According to the NRC, WASH-740 has been made obsolete by new discoveries and the consequences would be "less severe": an assurance that consoled nobody on the evening of 30 March.

People asked each other, "*How* remote is the risk?" They had begun to realise that, because a meltdown is an *uncontrolled* nuclear reaction, scientists and engineers are themselves ignorant about what it entails and how to handle it. MIT Professor Henry Kendall succinctly described the position of all nuclear "experts" at that time: "They are way out in an unknown land with a reactor whose instruments and controls were never designed to cope with this situation. They are like children playing in the woods."[3]

As the tension mounted that Friday, one became ever more acutely aware of the ease with which nuclear fission, the most brilliant and menacing of all man's discoveries, can outwit the human brain. At TMI it was doing its own thing and all the nuclear experts in Bethesda didn't know how to stop it. They rushed to and fro in helicopters, and consulted computers, and formed Bubble Squads, and conferred on hot-lines, and were bewildered and arrogant and frightened and aggressive and placatory and evasive — and always unsure . . .

Those days were hot and sticky at William's Corner, like heavy, cloudy July days at home — which my friends said was unusual. On the Friday afternoon, strolling by a little lake where water-turtles were basking on half-submerged tree trunks, I suddenly saw a disconcerting parallel between technological America and the basically stone-age culture of the Aztecs. The Aztecs' state was dominated by a pantheon of fearsome gods who controlled the cosmos and could, if angered or misunderstood, destroy the human race. The Americans' state is dominated by a pantheon of equally fearsome gods who, if misunderstood, could rather more effectively do the same thing. The Aztecs believed that they depended on their gods for protection against invaders and for the provision of food, light, heat and energy: some Americans are similarly deluded. The Aztecs were afraid to annoy any god lest the sun might cease to shine or the rain to fall or the crops to grow. Many American politicians have also been conditioned to believe that their idols — weapons and reactors — are essential for the safety, health, comfort and general well-being of society. The Aztec state was run by an élite of mathematical specialists; only these priests could make the necessary complicated astronomical calculations on which were based successful magic formulae. Therefore Aztec society was plagued by fear and anxiety: if one of the élite got his sums wrong, anything might happen. American society is heading the same way, towards a dependence on what is already known as "the nuclear priesthood". The Aztecs had to propitiate their gods regularly by offering them the still-palpitating hearts of human sacrifices. The Americans have not quite reached the human sacrifice stage but at TMI they approached it.

While I was thus gloomily brooding, Harold Denton was arriving at TMI, as promised by President Carter, with an entourage of some dozen "experts" from NRC headquarters. Earlier in the day, the NRC had learned of Wednesday's hydrogen explosion: a time-lag that strains credulity. But so it was. However, the NRC did already know that the reactor contained a gas bubble of some sort and by Friday morning it had been generally recognised that this bubble — an estimated 1,000 cubic feet of gases — was partly composed of hydrogen. Could it explode again? Louder and Worser? Denton worked all day on estimates given him before he left Bethesda, which suggested that the bubble could not ignite for five to eight days. His whole concern was to find a means of eliminating it in the time available. He was, however, working on *incorrect* estimates, which for almost three days caused the NRC to misunderstand the true nature of the TMI problem. A B&W employee, and certain NRC staff-members, had supplied correct information. But the providers of the official estimate ignored this and insisted that enough oxygen was being formed to cause a fire or explosion — which prompted the NRC, on Friday evening, to devise yet another set of evacuation proposals.

At 8.30 p.m., as we were dining peacefully at William's Corner, talking of books and forgetful of nukes, the unhappy Thornburgh (who was rapidly becoming one of my pin-up boys) had his first personal briefing from Denton. This cannot have given the Governor sweet dreams. Denton explained that core-damage was severe and that core-cooling was being inhibited by the Bubble. (By now it seemed to deserve a capital.) But he added that no immediate evacuation was necessary. He and Thornburgh then held their first joint press conference, at which the Governor lifted his ban on people within a ten-mile radius of TMI leaving their homes. However, he repeated his advice that no pregnant women or young children should approach within five miles of the plant.

That afternoon, the Middletown campus of the Pennsylvania State University had been closed for a week. And just before midnight the Pennsylvania Emergency Management Agency, spurred on by Denton's Bubble anxiety, unexpectedly announced twenty-mile-radius plans. Six counties were instructed to prepare for the possible evacuation of 650,000 people, thirteen hospitals and a prison. Had such a move proved necessary, what might have been the consequences? The President's Commission was "disturbed by the highly uneven quality of emergency plans and by the problems created by multiple jurisdictions [i.e., ABCD] in the case of a radiation emergency. We found almost total lack of detailed plans in the local communities around TMI." This was why the Governor was so reluctant to approve a mass-evacuation. He later explained: "There are known risks, I was told, in an evacuation. The movement of elderly persons, people in intensive-care units, babies in incubators, the simple traffic on the highways that results from even the best of an orderly evacuation, are going to exert a toll in lives and injuries. Moreover, this type of evacuation had never been carried out before on the face of the earth, and it is an evacuation that was quite different in kind and quality than one undertaken in time of flood or hurricane or tornado . . . When you talk about evacuating people within a five-mile radius of the site of a nuclear reactor, you must recognise that that will have ten-mile consequences, twenty-mile consequences, hundred-mile consequences, as we heard during the course of this event. That is to say, it is an event that people are not able to see, to hear, to taste, to smell . . ."

Already, by that Friday evening, between 50,000 and 60,000 voluntary evacuees had left the endangered area, causing some problems. Significant percentages of the staffs of hospitals, prisons and stores failed to report for duty — possibly by way of giving themselves apparently justified bonus holidays, rather than because they felt genuine fear. But of course thousands were *very* afraid — "shit-scared", as one young father said in a radio interview while packing up to leave his Middletown home, a few miles from TMI. That morning, half an hour after Molloy's radio warning of a probable evacuation, he had sent his wife and ten-month-old twins to stay with friends near New York City. "But I guess we're lucky," he added. "Not everyone has folk at a safe distance." By the following evening, Goldsboro, another small town near the Island, was left with only 10 per cent of its population.

So far the nuclear industry had remained extraordinarily pass-
ive, perhaps hoping its biggest crisis would go away if ignored.
But after James Floyd's little experiment with gas transference,
and while the evacuation muddle was revealing the "experts' "
subnormal IQ to the public, President Herman Dieckamp of
GPU, Met Ed's parent company, decided to assemble a high-
powered team to deal with the emergency. He and an aide
conferred with nuclear-industry leaders all over the country,
explaining the knowledge and skills required at TMI. Late next
afternoon, the first members of the Industry Advisory Group
(now we have ABCDG) arrived on site, met Dieckamp, identified
the most urgent tasks and decided who should tackle which.

Why had such a team not been assembled forty-eight hours
earlier, as soon as the Met Ed technicians knew of the core
exposure; which is, after all, the penultimate disaster in a nuclear
reactor? In a Supplemental View, at the end of the Kemeny
Report, Bruce Babbitt, Governor of Arizona, suggests part of the
answer. "Met Ed's operating license stems from the NRC policy
that no matter how small or unsophisticated the utility, it is
eventually entitled to wrap its arms around a nuclear reactor.
Nuclear technology continues to proliferate throughout the
industry, with some forty utilities now operating reactors and
many more waiting in the wings. There is no doubt that the
management quality of utilities varies much more than other
major industrial sectors, such as large chemical companies or
computer manufacturers. And because utilities are necessarily
monopolistic in nature, normal laws of competition do not apply;
badly managed utilities suffer financial problems but somehow
survive. It is now time to assess this situation and determine which
companies are qualified to handle such a technology ... It is
remarkable that this issue has not been previously confronted,
but it is again a product of the "accidents can't happen" syndrome."

The US Department of Health, Education and Welfare
(HEW) was not being afflicted by that particular syndrome. Its
officials were very worried indeed about a possible release of
iodine-131 from TMI and the lack of potassium iodide — a drug
to prevent radioactive iodine from lodging in the thyroid, where
it can cause cancer. (Glands saturated with potassium iodide

cannot absorb any more iodine.) No pharmaceutical or chemical company was then marketing sufficient medical-grade potassium iodide; but at 3 a.m. on Saturday morning the Malinckrodt Chemical Company agreed to provide 250,000 one-ounce doses. Working with Parke-Davis, and a New Jersey bottle-dropper manufacturer, they began a non-stop effort and at 1.30 a.m. on Sunday the first consignment reached Harrisburg. By 4 April, 237,013 bottles had been delivered and MacLeod was put in charge of its distribution. Yet another round of in-fighting started when Thomas Gerusky, head of the Bureau of Radiation Protection, requested some for his staff at TMI. MacLeod refused to issue it, arguing that if the public heard of any issue they would demand a general distribution. Thornburgh and Denton backed MacLeod, but Washington was on Gerusky's side. From the White House Jack Watson asked HEW to prepare recommendations for the drug's distribution and use. Those recommendations were sent to Thornburgh in a letter from the White House on the following Tuesday, but MacLeod continued to oppose a general distribution. By then the risk of high-level releases from TMI was much reduced and he feared that the potential adverse side-effects would in themselves cause a public health problem. He won; throughout the emergency the potassium iodide remained under armed guard in a warehouse. Then it was moved to Little Rock, Arkansas, for storage. It will no doubt be useful during the next nuclear crisis.

The Bubble hogged the headlines on 31 March — though media commentators could find little of substance to say about a phenomenon that was baffling both the Industry Advisory Group and the special Bubble Squad that had been formed at Bethesda.

If my gentle readers are still with me, they will by now have realised that ABCDG, and its innumerable offshoots, grievously exacerbated TMI-2's troubles. The whole episode afforded a microscopic view of a nation suffering from advanced dementia bureaucratia. And during the Bubble crisis this disease became acute.

Roger Mattson, Director of the Division of Systems Safety within the Office of Nuclear Reactor Regulation in the NRC, had spent most of Thursday and Friday pondering the novel problem of how to remove a gas bubble from a reactor. On the Friday evening, Hendrie asked him to find someone to figure out the *rate* at which oxygen was being generated and *when* a hydrogen explosion might happen — if it did. The fundamental fear was of radiolysis, a process during which radiation breaks apart water molecules, separating their hydrogen and oxygen. For the accumulated hydrogen to explode (or burn: a less terrifying possibility, though not soothing) enough oxygen would have to be released to form an explosive mixture.

Soon after 1 p.m. on the Saturday, as I was boarding my train to go to Washington DC for a few days, Mattson began to get answers to Hendrie's questions. Four independent outside sources, all renowned for their radiolysis analyses, were agreed that oxygen *was* being generated, though their estimates of how much varied. They were also agreed that it would be several days before there was a potentially combustible mix in the reactor.

At 3.27 p.m. Mattson met the NRC Commissioners and told them: "We were not underestimating the reactor coolant system explosion potential; that is, the estimate of two to three days before reaching flammability limit was a conservative one." Later, however, Mattson's analysts changed their minds and told him that the Bubble was "on the threshold of the flammability limit". By that time he had already held discussions about the likely results of a hydrogen explosion. One NRC consultant predicted the blast would produce pressures of 20,000 pounds to the square inch within the reactor. But representatives of its designer, B&W, allowed for the dampening effect of water vapour and guessed the total pressure might be about 3,000 to 4,000 lbs/sq. inch — which, according to analyses, the TMI-2 reactor should be able to withstand.

If anyone still believed in analyses, their faith must have been shaken when, soon after the "threshold" message from the radiolysis "experts", John Taylor of B&W firmly repeated what another B&W engineer told the NRC on Thursday night: that *no* excess oxygen was being generated so there could be *no* explosion. Mattson later denied ever having received this not unimportant message — perhaps it fell into the undergrowth somewhere in the ABCDG jungle. But he admitted, at a hearing of the President's Commission, that by Friday the NRC could themselves have determined, from the information then available to them, that there was no danger of an explosion.

Throughout Saturday afternoon and evening, Hendrie and the Bethesda NRC officials had worked diligently to keep Denton briefed on the oxygen estimates and explosion potential. Yet he didn't hear of the "threshold" theory — then being taken seriously by some NRC staff at headquarters — until the Associated Press (AP) broke the news (verified by Denton's deputy at Bethesda and an NRC Public Information spokesman) just before Denton went to Harrisburg to hold a late evening joint press conference with Thornburgh and Scranton.

At the time, even to think of the Bubble's alleged potential made one feel quite queasy; but in retrospect it can be seen that things had now reached the comic-opera stage. President Carter had announced earlier in the evening that he would visit TMI next day — so what were Denton and Co. to do? Predictably, Denton told his second-in-command at TMI, Victor Stello Jnr, to contact a fresh batch of outside "experts" to provide a further analysis of the oxygen-hydrogen situation. Stello moved fast; already a Presidential aide had rung him to enquire about the AP story and he had emphasised that nobody at TMI shared the concern felt at NRC headquarters. But perhaps they *should* have been sharing it? It would never do carelessly to sacrifice the President of the United States to a nuclear idol . . .

At that evening press conference, Thornburgh assured reporters: "There is no imminent catastrophic event foreseeable." And Denton backed him up: "There is no near-term danger at all." When reporters challenged Denton about the extent to which he and his Bethesda colleagues were contradicting each other, he resolutely affirmed: "No, there is no disagreement. I guess it is the way things get presented."

HEW was also busy that day. In the morning, senior health officials gathered to discuss evacuation and concluded that an *immediate* evacuation, rather than a Presidential visit, should take place if the NRC could not guarantee by early morning that the reactor was cooling safely. The HEW secretary sent a memorandum to that effect to Jack Watson of the Presidential staff. Later, HEW attended an ABCDG mass-meeting at the White

House, convened by Watson, and there they urged that NRC officials should consult them and Environmental Protection Agency (EPA) experts, about the potential health dangers associated with the struggle to subdue TMI-2. But at no stage did the NRC seek HEW's advice.

That afternoon my train to Washington ran for miles along the Susquehanna River, into which 120,000 gallons of radioactive water had by then been dumped. The *Washington Post* reported soothingly that: "Max Eisenberg, acting Director of Maryland's Department of Environmental Health Services, said yesterday that scientific checks of air, of the Susquehanna River water that flows into Maryland from Pennsylvania, and of milk delivered there from dairy farms near TMI 'show no increase above normal' in radiation levels." However, a week later, in Washington, a group of scientists, including Professor Karl Morgan of the School of Nuclear Engineering at Georgia Tech., held a press conference and accused the Government of "withholding information about the dangers of the TMI accident and understating the extent of damage to the environment". They warned that "contaminated fields and water constituted a continuing threat to residents in the Harrisburg area and others downstream along the Susquehanna River as far as Chesapeake Bay".

Since the first alarm there had been much concern about possible contamination of dairy products. The economy of the Pennsylvania Dutch countryside surrounding TMI is based on the dairy industry and milk is packaged by more than 1,000 food-processing plants. The area's largest processor, Hershey Foods, uses one million gallons a day. And most Americans are well aware that milk transmits iodine-131, if contaminated grass is eaten by lactating cows. But in the *New York Times* of 8 April Gordon MacLeod was reassuring; he described as "trivial" the levels of contamination so far found around TMI. Commenting on this, Richard D. Lyons wrote: "Exact present levels may be meaningless, because it is more than likely that the peak of radioactive contamination in milk has not yet been reached." I recalled this on 23 May, when the *Irish Times* published a photograph of white-coated men and sad-looking cows. The caption read: "Pennsylvania State agricultural officials examining cows at the

Clair Hoover farm in Bainbridge, where twelve calves and seven cows have died since 5 April. State officials deny any connection between the deaths and the accident at the nuclear power station at nearby Three Mile Island."

I had bought a stack of newspapers for the journey and these reflected the public's apprehensive bewilderment. The *Evening Sun* (of Baltimore, Maryland) ran dramatic headlines: GAS BLED FROM REACTOR ROOM IN EFFORT TO AVOID MELTDOWN: TOUCHY PROCESS INCREASES CHANCE OF EXPLOSION. At one press conference, Jack Herbein of Met Ed had explained that the perilous shutdown procedure had already begun and overnight the gas bubble had decreased in size by one-third. Soon after Harold Denton of the NRC had announced that the bubble's size was "essentially unchanged" and he promised that no shutdown move would be made for several days, and then only after giving the public due warning. Met Ed had described degasification as "an agreed first step"; the NRC had described it as an "experiment". If the latter were right, then surely — I thought — a general evacuation *should* be taking place? (Which was precisely HEW's argument at the White House that same afternoon.) I said as much to the improbably tall man who was sitting beside me, having leg-disposal problems; his bald pate, grey goatee beard and pince-nez made him look ridiculously like an Edwardian inventor. As my anti-nuke badge was invisible, his reply could be uninhibited: "Governments can't afford to panic, they just have to ignore pressure from the press or way-out agitators. It's their job to look at the *whole* picture. There's a good chance all this will sort itself out — and what would be the nationwide effects of a mass-evacuation in a crisis atmosphere? The nuclear industry could be ruined. And we need it. It employs tens of thousands of our best brains. Our nuclear exports bring in billions of dollars a year. As it is, we already have deadly inflation and a desperate energy shortage and a crippling oil-bill. And the industry's critics are getting more powerful, or at least more noisy and troublesome and organised. It would be crazy to endanger it even more. Giving in to mobs is no way to run a country. Governments have to think of the greatest good of the greatest number."

I failed to take up the challenge; my weapon is the pen, not the spoken word. Instead, we discussed the Incas and the Mayas; Mr

Long-Legs proved, not surprisingly, to be an archaeologist. Then we turned to the Galápagos, and the peculiar nature of cats, and the even more peculiar nature of the Irish, North and South; on all of which topics we saw eye-to-eye, despite the intervening pince-nez.

That encounter set me thinking about the fundamental differences between pro- and anti-nukes. These seem to be a matter of temperament, rather than of principle. Or perhaps I should say a matter of vision, or adaptability. By temperament I am, after all, a fatalist and not disposed to fuss unduly about physical risks of the ordinary sort. But like all anti-nukes, I refuse to go along with a prevailing system merely because it exists. Pro-nukes, on the other hand, seem to feel that because a system has been *established*, by Someone Up There, it must be supported. Otherwise exports may fall, inflation may get worse, oil may run out, unemployment may increase — anarchy may prevail. Pro-nukes, then, are essentially timid, Aztec-types who have been sold the idea that nuke idols are necessary to preserve their way of life. They are not *ipso facto* any less sensitive, thoughtful or concerned about others than anti-nukes: but they are either afraid of or cannot imagine the unknown – a New Society.

Around Washington that weekend, the ethics of evacuation planning — or the lack of it — were being much discussed. Most people agreed on the need for a streamlined, hierarchical, generally understood evacuation plan, to be practised regularly by people living within a twenty-mile radius of nuclear power plants. The snag is that such a plan, practised often enough to work smoothly, savours of wartime conditions and would inevitably antagonise the public — apart from arousing widespread fear and suspicion of an industry that makes such extreme precautions necessary. Yet I heard many Americans, not all of them anti-nuke, arguing strongly that any government which allows the building of potentially lethal power stations is morally bound to insist upon minutely planned evacuation procedures and compulsory drill.

An alternative to such unpopular preparation for disaster has been suggested by Alvin Weinberg, one of the original developers of commerical nuclear power.[4] He feels it might be wiser to concentrate many nuclear plants — run by *one* organisation, independent of ABCDG— in a few remote, high-security areas, together with reprocessing plants and waste storage facilities. This would reduce the risk of perilous TMI-type *contretemps,* and of transport accidents and thefts. Also, it would require fewer armed guards than are at present employed in nuclear grave-yards and would eliminate the need for mass-evacuations. But it would also create most alluring military targets and terrorist challenges. And, within these hyper-protected, isolated, sus-picion-breeding fortresses, the "nuclear priesthood" could not be expected to remain normal for very long.

At this stage Met Ed and the NRC were sharing the media's angry scorn. Richard Pollock, a Washington anti-nuke leader, was widely quoted: "The NRC let this plant go forward despite the red flags that were waving. They look, but they do not see." The Kemeny Report was to endorse Mr Pollock's view: "Two of the most important activities of the NRC are its licensing function and its inspection and enforcement activities. We found serious inadequacies in both. In the licensing process, applicants are only *required* to analyse 'single-failure' accidents. They are not re-quired to analyse what happens when two systems fail independ-ently of each other, such as the event that took place at TMI . . . Plants can receive an operating licence with several safety issues still unresolvéd. This places such a plant in a regulatory 'limbo', with jurisdiction divided between two different offices within the NRC. TMI-2 was in this status at the time of the accident, thirteen months after it received its operating licence."

By 1 April I had realised that the silver lining to radioactive clouds is nuclear jargon. For all the ominous talk on the wireless about meltdowns and hydrogen bubbles, I couldn't help giggling when one expert referred to "a rapid disassembly". He meant, of course, an explosion . . .

During Saturday night and Sunday, ABCDG was brought to its collective knees by varying degrees of suspense, frustration, resentment and exhaustion. Every relevant office was plagued by incessant telephone calls from distraught citizens demanding to know which of the conflicting reports about the Bubble was *true.* One report had rather incoherently explained that to reduce the pressure in the reactor might enlarge the Bubble and uncover the fuel rods thus causing a meltdown. This inspired some bizarre offers; several heroic (or nutty) citizens were anxious to save Pennsylvania and/or acquire immortality by entering the reactor and releasing the hydrogen by hand, as it were. But, even had a safety valve existed for this purpose, no living creature could have survived the radiation in the containment room for long enough to reach it.

Official frustration was particularly acute in Dauphin County because Denton had cut off the flow of information to local level. Oran Henderson — Director, as you won't remember, of the Pennsylvania Emergency Management Agency — was now excluded from the Governor's briefings and press conferences. And when, in an attempt to right this, the State Senator George Gekas rang Thornburgh shortly before midnight on Saturday, he was told that both the Governor and Lieutenant Governor Scranton were too busy to talk to him. Gekas then threatened that Dauphin County would order an evacuation at 9 a.m. on Sunday, unless given more information and cooperation. At 2 a.m. Scranton rang the county's emergency centre and agreed to meet officials later in the morning . . . He arrived at 10 a.m., to find Henderson ahead of him, rampaging about his own lack of information. Scranton listened attentively to Molloy and his staff. Later, Molloy said: "I think he was just totally shocked by what was transpiring at our level; how busy we were; how much work we were doing; how complicated it was."

By that time, the latest team of "experts" had convinced Stello that there was no danger of an explosion. He explained to Denton that "pressurised water reactors, the type used at TMI-2, normally operate with some free hydrogen in the reactor coolant. This hydrogen joins with the oxygen freed by radiolysis to form another water molecule, which prevents the build-up of oxygen to a quantity that would allow an explosion to take place." On 1 April the TMI reactor could scarcely be described as operating "normally"; yet to take this into account, and cancel the Presidential visit, would certainly provoke an unmanageable, spontaneous mass-evacuation.

A few moments before the President was due to arrive at Harrisburg Airport, at 1 p.m., Hendrie and Mattson met Denton and Stello in a hangar. Mattson and Stello had not communicated for forty-eight hours. Mattson outlined the "threshold" theory, then being officially supported at NRC headquarters. In evidence to the President's Commission, he described the reaction: "And Stello tells me I am crazy, that he doesn't believe it, that he thinks we've made an error in the rate of calculation . . . Stello says we're nuts and poor Harold [Denton] is there, he's got to meet the President in five minutes and tell it like it is. And here he is. His two experts are not together. One [Mattson himself] comes armed to the teeth with all these national laboratories and Navy reactor people and highfalutin PhDs around the country, saying this is what it is and this is his best summary. And his other [Stello] saying, 'I don't believe it. I can't prove it yet, but I don't believe it. I think it's wrong'."

When the President arrived, Denton came clean and "told it like it was". Undeterred, Mr and Mrs Carter went straight to TMI-2 (lesser spirits like myself would have gone straight back home) and gallantly toured the plant. That evening, press photographs depicted a plainly terrified President walking through the control room, showing the whites of his eyes. (Mr Carter is himself a nuclear engineer.) Afterwards, the President warned the locals that, though radiation levels near the plant were now "quite safe", officials had to take critical decisions during the next few days; he implied that a general evacuation might yet be necessary.

Meanwhile, still more experts, including those at Westinghouse and General Electric, were being consulted. According to Mattson: "By three o'clock, we're convinced we've got it. It's not going to go boom!" The NRC Bethesda "experts" eventually crawled to the same conclusion, much later in the day. At 4 p.m., three NRC Commissioners were still seriously considering a general evacuation, while at the same time new measurements taken at TMI showed that the big Bubble was shrinking. Although the gases were still present, they had divided into smaller bubbles and this eased the task of eliminating the predominantly hydrogen mixture. Even now, no one knows why this happened. It was not because of any deliberate action taken by Met Ed or NRC engineers.

The NRC was entirely responsible for the public's unnecessary Bubble anxiety, yet they never announced, "It's not going to go boom!" Even after their own experts had reached that conclusion, by 6 p.m. on Sunday, Governor Thornburgh was not informed. As a result, most people remain unaware to this day that the possibility of a hydrogen explosion *never existed*. They also remain unaware of how close the plant came to meltdown, because the full estimated extent of core damage was first revealed only six months later by the Kemeny Report— a document not closely perused by the majority of US citizens, though in fact it makes fascinating reading if you have strong nerves.

Late that night Thornburgh asked all State officials to conduct business as usual on the following day, which people rightly took to mean that the evacuation threat was receding. But schools were to remain closed and pregnant women and young children were to stay well away from TMI.

I spent 1 April sightseeing in Washington and among the sights I saw was the Presidential helicopter returning from TMI — and then the President himself, doggedly jogging around the grounds of the White House. Heavily armed security men lurked in the undergrowth, occasionally emerging to give the few onlookers, straggling along the pavement, value for their tax money. To me it seemed that an assassin in a vehicle could very easily have picked off the President as he jogged close to the street; perhaps too long in Belfast has overheated my imagination.

Later, we heard that the sparse remaining population around TMI had been greatly heartened by the Carters' visit. A sixteen-year-old youth was quoted: "The President of the United States doesn't just walk into a danger area without knowing what is going on. It kind of makes you feel confident." Professor Ernest Sternglass was more sceptical: "I don't believe that even the President and his wife were told the actual dose levels; he may well have gotten more than the one millirem he was promised." And an eighteen-year-old Washington youth, with whom I shared my newspaper, said scornfully: "Carter'd risk his balls to save the face of the nuclear industry— and that's just what he has done today. Right?" "Right!" I agreed, hoping the relevant part of the Presidential anatomy would in fact come to no harm. Professor Sternglass was probably correct; three-and-a-half months later, the NRC admitted that at the time of the accident

TMI's steam release contained twenty-one times more radio-active xenon than had been expected by the experts.

The most worrying trait common to those "experts" is their inflexible obtuseness. In his Supplemental View at the end of the Kemeny Report, Commissioner Thomas Pigford, Professor of Nuclear Engineering at the University of California, Berkeley, made it plain that he disagreed with his fellow-commissioners' attitude to the problems of the NRC, or rather, the problem that *is* the NRC. He pointed out: "The NRC and the nuclear industry have taken and are taking steps on a broad basis to analyse and rectify these problems, as evidenced by the post-TMI NRC bulletins and by the establishment of the utilities' Institute for Nuclear Power Operations (INPO) and the reinsurance pro-gramme." How could anyone of even minimal intelligence approve the creation of yet another layer of nuclear bureaucracy? Unless INPO is to *replace* ABCDG, one can only see it causing more and worse problems.

Professor Pigford's comments on the effects of a meltdown are equally alarming: "Our staff analyses show that even if . . . a meltdown had occurred, the containment would still have survived and protected the public." Yet the Report itself stated bluntly: "At this stage we approach the limits of our engineering knowledge of the interactions of molten fuel, concrete, steel and water, and even the best available calculations have a degree of uncertainty associated with them." As long as nuclear "experts" continue to believe blindly in their demonstrably fallible analyses, the whole world is at risk.

Early on 2 April, a meeting of NRC officials decided that the tone of their imminent "Bubble" press conference must be ambiguous. Mattson explained to the President's Commission: "We wanted to go slow on saying it was good news. We wanted to say it is good news, do not panic, we think we have got it under control, things look better, but we did not want to firmly and finally conclude that there was no problem. We had to save some wiggle room in order to preserve credibility. That was our judgement."

Before Denton and Mattson met the press, a Met Ed official, George Troffer, had already bluntly told a reporter that the bubble was almost gone. But, though Denton admitted a "dramatic decrease in bubble size", he added that more sophisticated analyses were needed "to be sure that the equations that are used to calculate bubble size properly include all effects". On the explosion scare, he said smoothly, "The oxygen generation race that I was assuming yesterday when I was reporting on the potential detonation inside the vessel is, it now appears, seen to have been too conservative." He repeatedly referred to the NRC estimates as having been "too conservative" and adroitly hid their monumental bloomer. After such a display of duplicity, how can the general public be expected to take the NRC's word on any subject?

On 3 April GPU, Met Ed's parent company, appointed Robert Arnold, a vice-president of another of their subsidiaries, to organise the formidable and perilous clean-up of TMI-2. Next day, schools beyond the five-mile radius were reopened; gradually the afflicted area was recovering its equilibrium. But it is unlikely to return to normal for quite some time; the Kemeny Report noted: "The accident at TMI did not end with the breaking-up of the bubble, nor did the threat to the health and safety of the workers and the community suddenly disappear." For days technicians continued their struggles to bring about a "cold-shutdown" — achieved when the temperature of a reactor's coolant falls below the boiling point of water. Then they were left to ponder more than a million gallons of radioactive water — 690,000 gallons of it *highly* radioactive — plus immeasurable quantities of radioactive gases and a wrecked and still menacingly radioactive core. Several auxiliary buildings also had heavily contaminated equipment, walls and floors. An unprecedented decontamination effort lay ahead: unprecedented in its duration, cost and risk. When early clean-up attempts were made, in April and May, workers had to wear special clothing and respirators (despite which they received abnormal doses), as they mopped, wiped and vacuum-cleaned. At the end of August five workers had received doses far in excess of the NRC's quarterly limits for exposure to the skin or the extremities. Within six months, twelve truckloads of solid, slightly radioactive waste — clothing, rags, ion-exchanging resins, swipes, contaminated air filters — had been removed to the ironically named Richland, Washington, for burial at a commercial disposal site. And the worst of the job, technically and politically, still lay ahead.

By this stage Governor Thornburgh and most Pennsylvania officials were openly hostile to Met Ed. In a ten-minute television interview on 6 April the Governor had denounced them for having ignored or loosely enforced government safety standards and he recommended that "coal, wood and sunshine replace nuclear energy as a future source of power". Of TMI he said, "I have serious doubts as to whether the plant should ever be opened again." And, because the nuclear industry knew that Governor Thornburgh spoke for many new recruits to the anti-nuke cause, they dared not publicly do anything risky to expedite the TMI clean-up. Yet they had to seek permission from the NRC to release controlled bursts of krypton-85 over a two-month period, as radioactive gases were preventing entry to the containment building. The NRC, aware now of its every move being critically scrutinised, hesitated, and went on hesitating, for months. Then, during mid-February 1980, there were two radiation leaks from TMI-2; a pump was misbehaving. These leaks made a total of four radiation "mishaps" in the US within three days. But the NRC, true to form, said cheerfully, "Tests indicate that no radiation has escaped into the atmosphere."

The two leaks, which had been serious enough to require worker evacuation, brought Victor Stello, now Chief of NRC E&I, back on stage. Having visited the Island, he reported that the first leak, of between 700 and 1,000 gallons of radioactive water, was "well handled by plant personnel". But the second, which occurred during an air-sampling operation, was muffed. (This was unwontedly harsh talk from an NRC E&I official, but perhaps Mr Stello had studied the Kemeny Report: "The NRC is vulnerable to the charge that it is heavily equipment-oriented rather than people-oriented. Evidence for this exists in the weak and under-staffed branch of NRC that monitors operator training and in the fact that inspectors who investigate accidents concentrate on what went wrong with the equipment and not on what operators may have done incorrectly . . .") Stello concluded with a warning that there must inevitably be "very slight radiation releases" while so much highly radioactive water and gas remained in the containment.

The NRC then ordered "a quick study" to determine whether the clean-up could or should be hastened. One of the Commissioners, Victor Gilinsky, made a statement that my patient readers will at once recognise as classic NRC-speak: "The bottom line has got to be protecting the public. This may mean moving faster, or it may mean moving slower, but I think we've got to take a hard look at it."

On 6 March 1980, NRC "experts" mentioned the possible danger of a chain reaction restarting in the still-hot core. This risk could be dealt with only after engineers had entered the containment; so the NRC was prepared to sanction the release of 44,000 curies of krypton-85. However, on their releasing a tiny amount a few days previously the State authorities and local residents had "gone spare". One of many indignant letters in the *Middletown Press and Journal* complained: "We have a new modern version of dictatorship, aggression and human atrocities. The corporate officials of TMI are dictating the venting of voluminous radio-active poisonous gases over a period of sixty days or more which is against the will of the vast majority of people in this area." The writer was probably exaggerating the danger of the release in question; but then, where low-level radiation is concerned, nobody knows . . . Joseph Hendrie found these protests irritating. He said: "We can't sit around here and calculate environmental impact while we get ready to have a disaster." (Hendrie was still a commissioner, though Mr Carter had demoted him from the chairmanship of the NRC after reading the Kemeny Report in which he was accused of underestimating the importance of the TMI radiation leaks. In March 1981 Hendrie was reappointed Chairman by President Reagan.)

A fortnight later the NRC were challenged by a noisy, angry group of locals, among which the females were much deadlier than the males. (Perhaps they always are in the US.) One woman yelled: "We'll never forgive or forget what you've put us through. You're no more worthy than a hunk of cow manure." Another shouted "We hate your guts!" at a biologist. In defence of the NRC, Met Ed argued that the only other way to free the gas would be to build a special container for it, at vast expense. This would take at least two years, during which anything could happen in a reactor not designed to run for years (or even months) without maintenance. And the complex transfer process, never previously attempted, might itself be more dangerous than releasing krypton-85. Even if it were successful, the gas would merely have been moved, not discharged and forgotten about. The locals retorted that, having created the lethal stuff, Met Ed should be forced to accept full responsibility instead of dispersing it, at no cost to themselves, as an addition to our planet's accumulating envelope of man-made, low-level radiation. One can understand their fury, but in this case the gas-release may have been the *least* dangerous course to take.

America, whether irradiated or not, remains the Land of Opportunity. Within weeks of the accident, those obscure townlets all around TMI, which had been so violently dragged into the global limelight, realised that their misfortune could almost effortlessly generate bucks. Soon grisly "near-disaster" souvenirs, in rather doubtful taste, were being briskly flogged to thousands of morbid tourists. At Harrisburg Airport the gift-shop stocked up on note-pads and calendars featuring the world-infamous cooling-towers, and displayed money-boxes inscribed "Canned Radiation". At a souvenir shop only a few yards from TMI, near Middletown, lampshades were printed with blurred versions of the plant, sprawling sinister on its island. And T-shirts said, "Squeeze me, I radiate. Kiss me, I melt down." Or, "Happiness is a cool reactor." During the six months following the accident, Met Ed's heartily welcoming visitors' centre, overlooking the much-abused Susquehanna, had 50,000 sightseers, as compared to 10,000 during the previous twelve months. These were shown a twenty-minute film giving Met Ed's version of the accident.

To be able to laugh at one's own ill-luck is an excellent thing, yet many people find these jokey TMI T-shirts unamusing. Perhaps this is because TMI was not in fact merely a "near-disaster" as the nuclear industry would have us believe. For the victims of low-level radiation — one thinks especially of Met Ed's employees — it will prove, in time, to have been a tragedy. By April 1980 it was being widely reported that in the first six months after the accident the infant death-rate more than doubled within a ten-mile radius of TMI. Between April and September 1979, thirty-one babies below the age of one year died. In the same period in 1978, fourteen died. And thyroid abnormalities are rising above the national average. Gordon MacLeod, who in October 1979 was dismissed from his post as Pennsylvania Secretary of Health and is now a Professor at Pittsburgh University, has said: "The infant deaths were not necessarily caused by radiation from the plant — an alternative reason could be the psychological stress which the accident provoked in pregnant women and the mothers of young children. But it is important for the causes of the deaths to be investigated."

On 4 April 1980 the fires of local anger were further stoked by the publication of the seventy-page report on TMI put together by the Energy Subcommittee of the Science and Technology Committee of the House of Representatives. This blamed the press for causing "public hysteria" about the venting of krypton-85 from TMI and the Chairman, Mr McCormack, stated categorically: "This gas is harmless." However, two other committee members pointed out that "the contents of the report depended almost exclusively on the testimony of nuclear energy supporters".

On 29 June 1980, after a legal attempt to stop the release had failed, hundreds of families evacuated themselves from the vicinity of TMI and the long-planned venting at last began. But four minutes later it was halted, when the radiation monitor alarm went off. The NRC hurriedly proclaimed "No danger!" and explained that in a few hours they hoped to be able to resume venting. It was then discovered that the radiation monitors were faulty. Harold Denton, smooth as ever, said: "The monitor had seen krypton and thought it was particulates." Given his record, it is scarcely surprising that a lot of people retorted: "Maybe it *was* particulates!" Then a different alarm system was used, which indicated "no problem" while 43,000 curies of radioactive gas were released into the atmosphere. Was this the quintessential technical fix? If one bit of technology is disobliging, try another . . .

On 27 July two Met Ed workers, William Behrle, aged thirty-six, and Michael Benson, aged twenty-seven, volunteered to become the first people in sixteen months to enter the TMI-2 reactor vessel. Wearing radiation suits and equipped with a two-way radio, they remained inside for twenty minutes, taking photographs and gathering radiation samples on swabs. Afterwards a Met Ed spokesman, David Klucsik, said: "Everything went according to plan. Later we expect a report on conditions inside the reactor and on the engineers' health."

We haven't since heard about the engineers' health, but that reactor report was shattering. On 8 August, Met Ed officials announced that the clean-up would cost some $855 million, double the original estimate, and take twice as long as had been expected; it could not be completed before November 1985. A decontamination supervisor said, "The most difficult job will be dismantling the broken core — it's filled with radioactive rubble."

Herman Dieckamp, the GPU Chairman, admitted that his company was in "a tenuous financial situation" because of the accident; but he believed public utility commissions would help it to avoid bankruptcy. He estimated the total revised cost at about $2,800 million, including the cost of replacing electricity lost at TMI-2. He added: "A levy on all US customers who use electricity generated in nuclear plants would produce a fund of $100 million a year to help GPU and other nuclear companies in similar predicaments." Up to that date, Met Ed's customers in Pennsylvania and New Jersey had already paid $263 million extra because of the accident and $257.4 million more was due from them, to pay for outside supplies.

Two years after the accident, neither the containment building nor the damaged core had been cleaned up. And some 700,000 gallons of radioactive water still lay eight-and-a-half feet deep in the building beside the reactor core. However, Met Ed announced on 20 March 1981 that they had at last devised a system to process the water, though the NRC had not yet approved its use. By then it had been admitted that the clean-up would take *nine* years and experts have given up estimating the final cost. It is unsurprising that no new orders for nuclear power plants have been placed in the US since 28 March 1979. On 21 March 1981 Herman Hill, head of General Electric's power systems division, announced: "If we get no further orders, the industry has only three or four years to live."

Many other countries have also got the TMI message. In Japan, construction work on most nuclear sites has made little progress since 1979, though the government had hoped to have fourteen new plants on stream by 1985. Early in 1981, the Mayor of the small Japanese town of Kubokawa became the first Japanese local-government head to lose office because of his pro-nuclear attitudes. And he blamed TMI for his defeat: "Since then people are alarmed by the leakages of radiation." The rabidly pro-nuke Suzuki government later expressed unease lest the success of the Kubokawa anti-nukes might encourage more vigorous campaigns throughout Japan.

The nuclear industry will not of course be easily defeated; and its allies are many and powerful. On 19 February 1981 nuclear safety specialists of the twenty-four-nation Organisation for Economic Cooperation and Development (OECD) published a report[6] stressing that human failures and not plant defects were the main cause of the TMI accident. They conceded that statements had been received mentioning "possible deficiencies in some components and the need to consider the interrelations between all the systems in the plant". But the gist of their report was emphatically reassuring: nukes are safe if properly operated. An argument which side-steps the issue of humanity's capacity for "eternal vigilance".

Russell Schweickart, a former Apollo-9 space-hero and in 1979 Science Adviser to California's Governor Brown, was interviewed six weeks after the TMI accident by Stewart Brand of the *Co-Evolution Quarterly*. He had by then closely studied the entire transcript of the NRC's deliberations during the crisis. He said: "The frightening thing is that the hydrogen explosion was not anticipated, that there was no instrumentation to monitor the build-up of hydrogen in the containment so that perhaps warning could have been given to the authorities responsible if in fact the public had needed evacuation and protection. That is an incredible condemnation of the design and instrumentation of these power plants . . . It's not that you have villainy here. You just have lack of foresight in the design process. One must assume that there will be accidents that we haven't foreseen, that we haven't planned for, that we could not preclude from occurring, and then the next step is to ensure that you have adequate control instrumentation — that is, visibility into the guts of the system — so that if an accident — *when* an accident occurs, you can at least have a chance of controlling it . . . I don't want to hold NASA up as the great standard, but it certainly dealt very seriously with that latter part — handling the unforeseen — of the design philosophy. Only something of the order of 20 per cent or so of the time we spent in simulators and trainers was devoted to learning how to fly the mission in situations where everything was working correctly."

Commenting on this, Stewart Brand wrote: "It is interesting to compare NASA's safety precautions (then) and the nuclear industry's (still, ten years later), and note that NASA endangered only a few astronauts and some national prestige whereas nuclear plants put hundreds of thousands of lives at risk as well as the whole national economy."

Schweickart also commented: "You could say, 'But each time we have one of these things we learn a tremendous amount.' Except if you look at the record it's not clear exactly how much we *do* learn . . . In most cases you're dealing with private utilities. People are concerned about the rates they have to pay for electricity, so there's always a legitimate desire to keep the rates as low as possible . . . If you're in the profit-making business, you're always trying to get away with the best product for the least cost — that's the famous bottom line. But the question is, where is the line that says, 'adequate for safety of public'?"

Rusty Schweickart describes himself as "the accumulation of my experience, and a lot of that experience comes out of a technical background, so my general orientation to things is that you don't make hasty decisions. You don't jump to conclusions, you don't get swayed by emotional things — you consider them, but you try to consider them in some rational fashion and put a weighting factor on them and add them into all the rest of the stuff . . . I find myself now with the scale gradually moving in the direction against continuing use and development of nuclear power . . . Moving towards elimination of nukes in some way, but not in a precipitous irresponsible way."

3

Econuts and Loose Screws

Some think this whole consciousness revolution is a sign of
mental decay . . . The opposite view is that we are witnessing
nothing less than the beginning of a mutation in human consciousness which is mankind's only real hope of survival.
                                        JOHN WREN-LEWIS

A favourite pro-nuke tactic is to label anti-nukes "irresponsible"
and "ignorant". One nuclear scientist has remarked: "It must be
recognised that the anti-nuclear movement is really an umbrella-
organisation for idealists of all stripes."[1] (Just occasionally I regret
my lack of artistic talent: this statement inspires a wonderful
vision of starry-eyed zebras marching resolutely under a gigantic
umbrella.) True, when some newsworthy event has been organised by anti-nukes, numerous irrelevant groups do leap on the
bandwagon, often to the detriment of The Cause. In August
1979, at the Carnsore Point demonstration in Co. Wexford, I was
aghast to find myself surrounded by Women's Libbers, IRA
representatives, Abortion For All, Hari Krishna and Co., the
Communist Party of Ireland and sundry other enthusiasts for
whom I feel little or no sympathy. In a rigidly conservative society
like Ireland's such hangers-on make it more difficult for the
embryonic anti-nuke movement to gain support. At Carnsore I
had reluctantly to agree with the "enemy" argument that "the
anti-nuclear movement has become the lightning-rod which
attracts all dissident points of view".[2] But happily the scene
is different in the US. There, too, the anti-nuke movement
inevitably attracts dissidents, yet within the past decade it has
largely overcome the handicap of a "way-out" image.

In San Francisco, on my way to Pennsylvania, I had visited

several anti-nuke groups in their offices and seen what is involved in being an effective opponent of nuclear power and nuclear weapons. It is easy to dismiss protest marches, sit-ins and demonstrations as mere diversions to wile away an idle hour for restless students. Yet such activities, when organised according to the principles of non-violence, require an immense amount of hard work, self-restraint, endurance and, at times, physical courage. This San Francisco impression was confirmed when I travelled on from Washington to Boston.

I had met Roger, my thirty-four-year-old Boston host, in the Ecuadorean valley of Otavalo, high in the Andes. There we talked for hours about the future of our planet, sitting in the patio of a tiny doss-house, and Roger invited me to continue the conversation in Cambridge, Mass., where he worked as a bicycle mechanic. But when I arrived in Peach St on 3 April 1979, in the wake of TMI, my host had no time to be sociable; as one of the leaders of the local Clamshell Alliance (an association of anti-nuke groups) he was busy helping to organise the next day's march to Boston Common.

Roger and several friends, male and female, lived in four two-storey nineteenth-century houses in a quiet street, ten minutes walk through Harvard from the subway station. And very remarkable those houses were: each had been rescued from collapse by its present occupants, then fully and comfortably furnished off the local municipal dump. Cookers, refrigerators, washing-machines, lamps, beds, tables, chairs, carpets, cupboards, lavatories, baths, electric kettles — even a toaster — all had been discarded because of some slight defect (soon fixed by Roger and Co.) or some insignificant blemish, like a cochineal stain on a carpet. The general impression was of homes several degrees less shabby than my own and with many more mod. cons.

Soon I was hurrying back to the centre of Boston, on Roger's advice, to see that eerily prophetic film, *The China Syndrome*, which had been released on 16 March. Shortly before that date, it was given considerable extra publicity — all for free — when the news leaked out that General Electric had been ill-advised enough to cancel their sponsorship of a Barbara Walters TV special which briefly mentioned the film. Twelve days after its release, TMI happened — the sort of coincidence every filmmaker must dream about. In fact, *The China Syndrome* was rather less dramatic than the reality. No radiation leaked into the atmosphere and there was no evacuation from the vicinity of the plant because the public remained unaware of the crisis until it was over. The makers had received excellent technical advice; unpopular as their product was with the nuclear industry, nobody could accuse it of scientific inaccuracy.

Most of the English-speaking world knows by now that *The China Syndrome* is a fast-moving thriller about safety cover-ups at a nuclear power plant. A potentially lethal accident is avoided by a nuclear engineer (Jack Lemmon) who then discovers that the plant's safety has been proved by the use of falsified X-rays of welds. A TV reporter (Jane Fonda), who is doing a documentary on nuclear power, happens to witness the near-accident and her cameraman (Michael Douglas) films it and is thus in a position to prove that the plant management is deceiving the public. Then the plot thickens hideously and is strongly influenced by the Silkwood tragedy. The utility bosses refuse to admit how close the plant has been to disaster and insist that such an "incident" could never recur. But it does: and a meltdown is averted only at the last moment.

To see *The China Syndrome* when I did was quite an uncanny experience; on the subway from Cambridge I had read in that day's *Boston Globe:* "Officials at Three Mile Island now say the likelihood of a meltdown is 'extremely remote'." Which meant still possible ... A strange frisson went through the audience when one of the characters explained that if the meltdown happened it would "render uninhabitable an area the size of Pennsylvania". Then slow clapping began. One felt that the lines between fact and fiction, reality and fantasy, had somehow been blurred: always a disconcerting sensation. A week earlier this film would have seemed to me just another eruption of sensationalism. This has been the most heartening world-wide consequence of TMI — what before seemed almost impossible, to nuclear experts and general public alike, is now known to be possible. And the stupidity, dishonesty and ruthlessness of the nuclear bosses and their allies, which would have overstrained the average cinema audience's credulity before TMI, has since appeared only too plausible.

Outside the cinema, keen young anti-nukes were distributing leaflets and urging people to attend the next day's march. They did not have to force their literature on the public; the audience queued for it.

On the subway I re-opened my *Boston Globe* and gloomily read an Associated Press report: "In Washington, Energy Secretary James R. Schlesinger said Congress should act to speed up licensing for nuclear power plants, despite the accident near Harrisburg, if the nation wants to reduce its heavy dependence on Mideast oil. In testimony at a House hearing yesterday, Schlesinger predicted that President Carter soon would resubmit to Congress a bill — unsuccessfully proposed in 1978 — to reduce from eleven years to about six years the time it now takes to approve, license and begin building a new nuclear plant . . . Meanwhile, seven nuclear plants in five states that use B & W reactors similar to that involved in the TMI accident were ordered by the NRC to run a safety check and report on their findings within ten days. A spokesman at B & W declined to comment on the order. In a statement over the weekend, the company denied any responsibility for the accident at TMI . . . A Public Utilities Commission of Ohio hearing-examiner turned down a request by Cleveland Mayor Dennis J. Kucinich to keep the local plant shut indefinitely. Kucinich cited evidence that the plant had been shut thirty-eight times since November 1977."

Back in Peach St a lot was going on, conversationally, and I listened. Until that evening I had not fully appreciated how much the manner of the Vietnam War's ending has affected America's anti-nukes. They feel that since public opinion stopped a major war, against the governmental will, an anti-nuclear campaign is not as impractical as it may seem. But, like all "movements", this one includes various groups and individuals with conflicting views about how best to achieve their common aim. To me — and to most of those in Roger's house that evening — the campaign *must* be nonviolent to succeed. Yet this is not easily organised. Mass-support — thousands marching, or obstructing the entrance to nuclear power plants, or protesting outside offices — tends to become over-emotional and unruly. Then the whole movement is brought into disrepute and the crowd becomes an easy prey to *agents provocateurs*. Therefore the theory and strategy of nonviolence have to be studied by anti-nuke leaders and the mass-support has somehow to be educated into following those leaders. All this was explained to me by one of Roger's friends, a twenty-eight-year-old house-painter whose father, assuredly, had not been a house-painter — any more than Roger's had been a bicycle mechanic. He concluded, "Through nonviolence people discover their social power." This no doubt is one reason why the authorities sometimes give in to nonviolent campaigns — lest more and more people might discover that by exercising their social power they can quietly disrupt the Establishment. Our society assumes that power is confined to governments, armies, institutions, corporations and extremely wealthy individuals. But, according to nonviolent theory, power, in a democracy, depends on the citizens' constant obedience. The rulers' power can therefore be shaken, and eventually demolished, by a refusal to obey the rules. Gandhi's civil disobedience campaigns first demonstrated this obvious yet usually ignored fact.

When nonviolent anti-nukes confront the police these two groups are not really in opposition but are competing for public support. If the anti-nukes seem well-behaved and amiable, the non-aligned public receives no threatening impression. Then, on seeing the police move in to arrest the demonstrators, people ask: "Why are these restrained and friendly people taking this sort of risk? And why are they being hauled off to jail?" The pro-nuke propaganda machine then has some trouble getting into gear. A Seabrook demonstration on 30 April 1977 drew over 2,500 supporters, from thirty-one States, and is now recognised as one of the best-organised nonviolent actions ever taken anywhere. Yet afterwards the pro-nukes referred to these demonstrators as "terrorists" — a vicious absurdity which only damaged their own case.

So conditioned are we to an acceptance — even approval — of violence, that the term "nonviolence" switches many people off. They associate it with waffling do-gooders, and somewhere there seems to be an implication of cowardice, or at least inglorious timidity — a shrinking from the rough-and-tumble of real life as it has been ever since Caveman Number One realised that he could get twice as much meat by hitting Caveman Number Two over the head. The nonviolent Christian sects, like Mennonites and

Anabaptists, have never made much impression on the rest of Christendom. And, though Gandhi was undeniably effective in India, that is another, and very different, cultural scene. More respect is accorded to the Quaker ideal of persuasion by example and education, rather than by coercion. But the Quakers' pacifism is not the sort of nonviolence we are now talking about. Their consistent self-effacement, their dislike for becoming involved in publicity and agitation, even today lead some of them to condemn as "violent in spirit" such nonviolent activities as strikes and boycotts, or the blocking of roadways when nuclear equipment is in transit.

The sort of nonviolence advocated by my Boston friends is, like war, a method of organising conflict. It demands of participants a readiness to take risks and endure suffering *without* retaliation. Five of that group in Roger's house had served prison sentences for taking part in the Seabrook civil disobedience action in 1977. Three of those had again been arrested, with 180 others, on 9 March 1979 when they attempted to obstruct the delivery of a 427-ton stainless-steel reactor pressure vessel. These were shortly to appear in court again, for arraignment on disorderly conduct charges, and two of them had been arrested five times, at various demonstrations. In my sort of essentially conservative world we tend to look upon persistent demonstrators with some suspicion. We ask — is there something unstable about them? Why are they willing to sacrifice themselves to this extent? Do they fancy themselves as martyrs? Are they craving excitement? Is there something important lacking in their lives? Are they cranky or exhibitionist or anarchic or just plain neurotic? Certainly Roger and his friends were conspicuously the reverse of neurotic. Ten or fifteen years ago they had been young hippies or "Flower People". Now they were calm, clear-thinking men and women with enough courage, optimism and determination not to take the easy way out by referring to "the powerlessness of the individual". Some had hostile families to contend with as a result of their total rejection of the way things are. Those who came from privileged home backgrounds could by this stage have been in well-paid jobs had they so chosen; but they were happier with what comes off the dump. I thought of them as the post-materialists. And they were fun to be with: not tedious fanatics but cheerful, relaxed, enthusiastic, well-balanced men and women who cared deeply about other human beings in an unsentimental, practical way.

This group did not judge the success of a march or sit-in by sheer numbers but by the behaviour of the demonstrators and the feeling that was generated. As one of them put it: "We want our actions to make people stop and *think*. Just as Martin Luther King's made them stop and think about Civil Rights. We don't want the whole nation to march with us, we only want them to be aware of the problem. And most of all we want them to realise that by exerting democratic pressures they *can* help to force the government to change its nuke policies. On the whole we get a good press. The media have gone cynical about pro-nuke spokesmen, so many inaccuracies have been exposed over the years. And after Vietnam and Watergate the public is pretty well prepared for any number of scandals to be dug up around the whole nuke industry — reactors *and* armaments."

Roger had just printed a leaflet giving the guidelines for the fourth occupation of the Seabrook site, planned for the following June. It said:

1 Everyone must receive preparation in nonviolent direct action before taking part
2 No weapons of any kind
3 No damage or destruction of Public Service Co. of New Hampshire or Seabrook property
4 No running, at any time
5 No strategic or tactical movement after dark
6 No breaking through police lines
7 No dogs
8 No drugs or alcohol
9 In case of confrontation, we will sit down
10 We will not block workers' access to the site
The preparations for such an action include a seven-hour training session in nonviolent procedures.

Another interesting piece of printed matter was *GREAT News* — *GREAT* standing for the Grass Roots Energy Advocacy Team. This tabloid has a convincingly grassroots look and is distributed all over the US. Nobody knows who finances it but it is published by the international public relations firm of Underwood Jordan Yulish Associates, which specialises in PR for the electric and nuclear industries. Its aim is to promote nuclear power and counteract the anti-nuke movement. *GREAT* offers — free — a hot-line for information, a media monitoring service, grants, expert resource assistance, a publications reprint service and "Access to continually updated directories of grassroots energy activists on all sides of the energy controversies". At the First National Energy Advocacy Conference, held in Washington DC in February 1979, Senator James A. McClure from Idaho described anti-nukes as belonging to a new class "which wants to limit growth so that no one else's enjoyment of the good of society will infringe on their own". He continued: "The supporter of nuclear energy must truly believe that nuclear energy is a moral necessity of mankind and that, without it, future generations will sink deeper and deeper into poverty and, eventually, dictatorship . . . Negotiation with nuclear foes is impossible because they are embarked on a quasi-religious crusade." Another speaker was more moderate, warning that: "We must not assume automatically that those who disagree with us are necessarily evil . . ." Yet the anti-nuke movement was repeatedly presented as a destructive force working to undermine the American way of life. Audiences were advised to "educate the public" along these lines, with the help of the Department of Energy and the Department of Education, who would be happy to provide funds and specialised assistance. All this delighted Roger and his friends; they saw the formation of *GREAT* as proof that America's anti-nukes have already significantly influenced public opinion; otherwise the government would not be opposing them with thought and money as well as jail and sneers.

I liked Senator McClure's diagnosis that anti-nukes are embarked on a quasi-religious crusade. This is certainly true of those who are prepared to be jailed repeatedly and have so arranged their lives that anti-nuke duties can always come first. They see this movement as but a part of their long-term campaign for a New Society. The case for nuclear power is based on the need to become independent of foreign oil, so that there can be no outside interference with the sacred American way of life. The case against it is based on a denial of the sacredness of the American way of life. Anti-nukes believe that independence of foreign oil may be achieved through *conservation* of energy. But can such a reformation come about — anywhere in the West and least of all in America — before we have experienced a massive disaster of some kind? We are at present in a trap from which the majority *do not wish* to be released.

The post-materialists talk a lot about entropy and in Roger's kitchen one young woman, called Anna, took a jar of instant coffee off a shelf and waved it at me. "Look at that!" said she. "We shouldn't buy it!" And she explained the intimate and fairly obvious connection between being anti-nuke and anti-High Technology foods.

To produce instant coffee you first build a large factory which does the environment no good, either aesthetically or chemically. Then coffee beans are roasted, ground and infused in enormous vats of very hot water. That liquid is poured into a series of huge stainless-steel evaporators where it is heated, by oil-fired steam, until only a sticky sediment remains. This residue is freeze-dried in gigantic refrigeration plants which send massive amounts of heat into the air and require heavy, electrically operated vacuum pumps to draw vapour away from the frozen materials. (By this stage one hesitates to describe it as "coffee".) A solid cake remains, to be crumbled, packed and sold. Still more fuel is consumed by the furnaces that make the glass jars. And the air is further polluted, and more oil is wasted, by the trucks that carry the empty jars to the factory and the other trucks that carry the full jars away. And trees are felled to make the labels that go around these jars ...

There is a pleasing illusion of *smallness* about Boston; everywhere one wants to go seems within easy walking distance and many of the natives actually do walk, leaving the twisting, one-way streets blessedly free of motor traffic. I had expected much of the city to *look* English; it fascinated me to discover that in places it also *feels* English. Outside Prescott's house I paused reverently to reflect upon the strange life of that blind scholar, living physically in a demure corner of Boston while mentally all involved in the hectic adventures and bloody deeds of Spaniards, Aztecs and Incas.

Stuffing numbed hands into my pockets, I turned up Fairfield St just before noon to join the pre-march rally at the Prudential Centre. About 3,000 anti-nukes had assembled — the vast majority youngsters, since this was a special mid-week march hurriedly organised by student groups. I seemed to be the only middle-aged protester. Flocks of home-made banners fluttered over the crowd and many marchers carried their own personal placards, usually with the simple slogan NO NUKES. Some of the banners said: GOODBYE HARRISBURG: WHO'S NEXT? Others demanded: ENERGY WITH HOPE NOT DREAD. Still others proclaimed: NUKES ARE THE RAPE OF THE EARTH. Most conspicuous of all was a grotesque eight-foot effigy, half-skeleton, half-witch, representing Nuclear Power, and dominating the whole scene.

This was an impressive rally, well disciplined and purposeful, serious yet good-humoured. A few policemen were visible in the distance, standing on the pavements, ready to reorganise the traffic when the march started. Nonviolent demonstrators always tell the authorities in advance exactly what they plan to do. This strategy is regarded as an extremely important part of their campaign, for several reasons. Any sort of furtiveness is considered inappropriate since nonviolent activists believe in the legitimacy of their actions and expect others to do the same. They also deprecate hostility to the police, as such, because individuals should never be victimised in retaliation against their employers. Moreover, nonviolent anti-nukes are passionately convinced that the New Society *must* have honesty as its foundation stone. Of course not everyone in the movement agrees with this. Some argue that because policemen are the willing agents of an unjust system they deserve no respect or cooperation. This sort of callow extremism naturally delights the pro-nukes; wherever it gains the ascendency over nonviolence it quickly undermines support for the anti-nuke cause.

At 12.15 p.m. the march moved off with almost military precision; stewards saw to it that we walked eight abreast and kept close enough together to allow single-lane traffic on the wide street. Our destination was the Governor's House on the far side of Boston Common; we were going to present Governor King with a symbolic one-way ticket to Harrisburg.

Beside me a tall, clean-shaven young man with horn-rimmed

spectacles carried a placard advocating SUNPOWER. I asked him if anti-nukes are always jailed or fined when arrested, or if juries sometimes acquit them. Usually, he said, they are found guilty, since the basis of their defence is "competing harms" — a law that provides for the breaking of a minor statute to prevent a greater public harm. Predictably, when the legislature has already decided that a nuclear power station is to be built, most juries are not encouraged to consider its potential harmfulness. However, on 30 January 1979, at the end of a week long trial, a jury at Waukegan, Illinois, deliberated for four hours and then acquitted twenty defendants accused of criminal trespass after the nonviolent blockading of the Zion power plant. This case is regarded as a legal landmark and a major triumph for the nonviolent approach. The state's chief witness, the head of Zion security, admitted that he had never felt there was any threat to his personal safety, the safety of his staff or the security of the plant. He conceded that weeks before the blockade the demonstrators gave the police an exact account of what they planned to do.

Most marchers were chanting in unison: *NO* NUKES! *NO* NUKES! *NO* NUKES! Or: TWO, FOUR, SIX, EIGHT, *WE* DON'T WANNA RADIATE! Or: WE *ALL* LIVE IN PENNSYLVANIA! Or: *PEOPLE*, NOT PROFITS! *PEOPLE,* NOT PROFITS! *PEOPLE* NOT PROFITS! The small crowds lining the pavements seemed, on the whole, well-disposed towards us. Only a few scowled or turned away; the majority smiled tolerantly, if not approvingly, and many office workers, chewing lunch-time sandwiches, leant out of high windows to cheer us on.

Two banners had thought-provoking inscriptions rather than snappy slogans. "In a world built on violence, one must be a revolutionary before one can be a pacifist." And, "Revolutionary changes can only occur through direct action by the rank and file, not by deals or reformist proposals." My friend was able to identify the first as a quote from a pioneer of American nonviolence, A. J. Muste, who died in the '60s as a very old man. Both statements seemed sensible to me; but they might easily be twisted by pro-nukes to "prove" that anti-nukes are Communist tools.

At Boston Common we were addressed by a radiologist, a lawyer, a physicist, a Senator and a Civil Rights leader who had worked with Martin Luther King. The feeling was such that the very words "Three Mile Island" provoked cheers with undertones of anger. A young woman beside me said, "Since it didn't melt down or blow up, it's the best thing that could have happened!" She was a Clamshell organiser and during the past week over 300 new members had joined her group. She told me that that morning she had spoken on the telephone to Bob Pollard, in Washington — an ex-NRC official who resigned for reasons of conscience. He had foreseen major problems with LWRs but nobody listened. And that morning he had said: "You don't have an ultimate catastrophe out of the blue. There's always some kind of warning. TMI was a warning." "So now," said the young woman, "it's up to us to make sure the warning is heeded." For me her attitude perfectly summed up the mood of that rally — urgency, determination, and a cautious brand of self-confidence.

When the speeches were over we all moved off to the Governor's House and thronged outside while two young men presented the one-way ticket to a staff-member. Half-a-dozen Capitol Police stood at the gates: burly, handsome, impassive. Now the crowd suddenly relaxed and began to sing and dance — no doubt partly to restore circulation — and there was an amount of horse-play. Several young men showed signs of wanting to bait the police but were firmly curbed by those around them. It was 4.30, and almost intolerably cold, before we dispersed.

On the way back to Cambridge I told Anna how much the march had impressed me; in America one can believe in an anti-nuke victory before the end of the century. But she frowned and shook her head. "We can't wait that long — it's a question of *investment*. If many more billions of dollars are poured into nukes, we'll have lost. We're at a cross-roads right now — much more so than countries further down the nuke track. Look at Sweden, Belgium and Switzerland — already 30 per cent dependent on nukes for electricity. And France even more so."

"Then what difference," I asked, "will it make if *you* decide against expanding the industry?"

Anna shrugged. "Well, we're the ones who've been pushing it all the way. And we're rich and powerful. Other nuke countries are still dependent on us for technology, not to mention loans. That's why there surely would be hell to pay, politically, in the short term, if we changed direction. So ordinary people every-where *must* be informed about the realities of the scene — then they'll feel relieved, not let down, if we pack it up."

The Peach St colony was preparing for a nearby Clamshell Alliance meeting at 7.30 p.m. As their local branch had just acquired hundreds of new members, there was much reorgan-ising to be done — including a certain amount of discreet weeding, since not all recruits would be temperamentally suited to nonviolent campaigns. I was invited along as an "observer" and, after the initial get-together in a huge hall, the meeting broke up into small groups and the old hands gave suitable recruits preliminary lessons in nonviolent techniques.

A young woman who had returned only the day before from Manila looked the reverse of nonviolent; she was carrying a bulky file and visibly seething with rage. In the Philippines, Westing-house, backed by the US and Philippine governments, is con-structing a $1.09 billion reactor nine miles from an active volcano in an area where earthquakes and tidal waves are common. No-body is even pretending that the Philippines *need* this plant; the electricity is to go to a nearby "free trade industry" of which 70 per cent is foreign (mainly American) owned, with legal repatri-ation of all profits. To make space for the reactor, some farmers have had their land requisitioned and others have had their crops destroyed by flooding. The local Atomic Energy Commission says no radioactive wastes can be stored in the Philippines but must be shipped elsewhere — possibly to the US, though it as yet has no civil reprocessing facility. Eleven thousand people live near the site and in Bataan province 25,000 people signed a petition against it. However, under martial law there can be no question of effective opposition to a government decision. The US Export-Import Bank is meeting $644 million of the cost, their biggest-ever loan. The reason why the government decided to build this plant has not been explained. Herminio Disini, a close relative of President Marcos's wife, owns the sub-contractor for the plant, Asia Industries, and three of the other firms poised to gain from the deal. Building started before the NRC had approved the shipping of the reactor to the site.

Of course nobody should be astonished by this decision to build a reactor in a far-off region of volcanoes, earthquakes and tidal waves; even in America the nuclear industry tries, with decreasing success, to take earthquakes in its stride. It calls them "seismic events", a euphemism presumably calculated to diminish public apprehension. And it emphasises the infrequency of these "events", as it happily constructs reactors in Class 3 earthquake zones — the zones of maximum risk. Yet it's not so long since 1899, when the ground slipped forty-six feet during the Yakutat Bay 'quake.

America's record of earthquake magnitudes has been kept for less than eighty years and it is in any case lunatic to pretend that scientists can precisely predict the force of future 'quakes — this is but another symptom of the arrogance of Technological Man. Yet such predictions do determine the strength of nuclear power plants built in earthquake zones. And the Atomic Safety and Licensing Appeals Board (ASLAB) does not always agree with the industry on the infrequency of 'quakes. In one report it notes: "The ability of nuclear power plants to withstand earthquake damage is undeniably crucial in California, where seismic phenomena are not uncommon."

Just before dawn on 17 January 1980 a minor earthquake tripped seismic instruments at the Indian Point nuclear power plant which serves New York City. Throughout the area windows were rattled and so were the residents. But a plant spokesman reported: "No damage."

Exactly a week later (seismic events are so infrequent!) hundreds were evacuated (though only twenty-four were injured, none badly) when a thirty-second 'quake, registering 5.5 on the Richter scale, was felt some forty miles south of San Francisco. This damaged the Lawrence Livermore Nuclear Weapons Research Laboratory (the equivalent of Britain's Aldermaston) where huge amounts of radioactive materials are stored. Jess Garberson, the public information officer, admitted that radioactive water was leaking "at the rate of several gallons an hour", from one of the half-dozen 30,000 gallon storage tanks that stand on stilts in a corner of the one-square-mile laboratory complex. But he insisted, "It is posing absolutely no health hazard to the public."[3] One wonders why the industry bothers to store these radioactive liquids in vastly expensive tanks, since leaks, whether in the US, the UK or Europe, are always said to be "harmless".

Next day Garberson dismissed press complaints about the leak as "a storm in a teacup". But he acknowledged serious damage to the buildings, though a stronger 'quake in August 1979 caused "virtually no damage".[4] Can the frequency of California's "seismic events" be wearing the buildings down?

According to Mr Garberson: "The water, with slight tritium [radioactive hydrogen] contamination is falling into a catch tank called a berm. It's a bit like a coffee cup spilling over into a deep saucer. The tritium content of the water is about half the concentration the NRC allows us to discharge into the sewers. It's almost clean enough to drink — it's more like used bath-water."[5] Perhaps after all Mr Garberson was slightly on edge; most Americans of my acquaintance would not be calmed by the thought of drinking bath-water.

During my stay in San Francisco, in early March 1979, many locals expressed concern about the Pacific Gas and Electric Company's (PG&E) plans to operate a two-plant nuclear power centre at Diablo Canyon, near San Luis Obispo. As I passed this site on the way north from Los Angeles our bus driver had referred to PG&E as "a greedy gang of homicidal maniacs" — seemingly intemperate language, yet forgivable in the context of a nuclear power station beside an earthquake fault. The driver's mother was living in San Luis Obispo and he wanted her to move if eventually the plants were allowed to start up — and that was *before* TMI . . .

The Diablo Canyon reactors have been built two-and-a-half miles from the newly discovered Hosgri fault — which PG&E failed to discover *before* building, despite warnings that it might be close. Even pro-nukes have been disturbed to realise that a fault, said by the US Geological Survey to be capable of delivering a 7.5 Richter-scale shock, has been found so close to reactors designed only to withstand, on their builders' admission, a 6.75 shock. San Francisco's 1906 'quake registered 8.2, which means that it was more than fifty times stronger than a 6.75 shock.

At the end of the '60s, when PG&E began work on this site, they assured an already slightly fretting public that the nearest major fault was forty miles away. At a public hearing a local geology professor mentioned the severe 'quake that shook the nearby Lompoc region in 1927 and urged PG&E to do some more seismic research before building. He was ignored — and soon after lost his job.

Then, in 1971, two Shell Oil geologists discovered the Hosgri Fault and at once strenuous efforts were made to halt building while its potential was being further investigated. The NRC refused to intervene but guaranteed that the plant would not open, even if it were completed, should the fault prove to be a major hazard. (At that stage many people asked, "What is a *minor* nuclear hazard?") An ASLAB comment went as follows: "This is more than a run-of-the-mill disagreement among experts. We have here a nuclear plant designed and built on one set of seismic assumptions, an intervening discovery that these assumptions underestimated the magnitude of potential earthquakes, a re-analysis and a *post hoc* conclusion that the plant is essentially satisfactory as it is — but on theoretical bases partly untested and previously unused for these purposes. We do not have to reach the merits of those findings to conclude that the circumstances surrounding the need to make them are exceptional in every sense of that word."[6] This last bit of gobblygook seems to mean that ASLAB had caught PG&E cooking the books.

By March 1979 fuel had been delivered to the Diablo Canyon site but was not yet loaded in the core; the NRC was still withholding an operating licence. However, Californian anti-nukes were pessimistic; it had been disclosed that in 1976 powerful Commission members believed that Diablo's licence must on no account be withheld "because of the large financial loss involved and the severe impact such action would have on the nuclear industry".[7] Also, Mr Shackleford, a PG&E director, had announced that he was confident of securing the licence by May 1979 because the $1.4 billion plant had been given new flying buttresses with steel-reinforced concrete and its instruments had been insulated to absorb greater shocks. ASLAB's comments, which had been published on 23 January 1979, were dismissed as unreasonably fussy; "like a hen with ducklings", said one PG&E public relations officer during a wireless interview.

From PG&E's point of view, the TMI drama could not have happened at a worse time. They had requested an operating licence in October 1973 and it seemed almost in their grasp, after a five-and-a-half year wait, when they were foiled by the TMI aftermath — which included a decision by the NRC to grant no more licences until their own inspection procedures had been "revised".

By 1980 PG&E's list of woes was considerable. (1) Routine safeguards not yet approved by the full Commission, i.e., the NRC Commissioners all voting together. (2) "Seismic" safeguards not yet approved by ASLAB. (3) TMI-related issues still being reviewed. (4) Emergency evacuation planning at a standstill. On that last point, rumour suggests that the industry, having been temporarily cowed by post-TMI disclosures, is reasserting itself behind the scenes. A San Francisco lawyer friend of mine has reported: "Essentially, at the federal level there appears at present to be a lot of duplication, numerous proposed regulations and criteria and very little concrete guidance. I have heard rumour but have not yet been able to confirm that some disturbing language related to emergency planning has been inserted into the NRC's 1981 FY Budget Authorisation Act. In the previous year's budget, Congress, reacting to TMI, had inserted language prohibiting the licensing and continued operation of nuclear plants unless states and locals had "concurred in" plans. Certain members of Congress apparently grew concerned that certain recalcitrant states (California) or Governors (Brown) or locals (San Luis Obispo) might refuse to prepare plans or might purposefully fail to promulgate plans simply to prohibit operation of new and existing plants. Thus, language was proposed and may have been inserted which permits the NRC to review and approve an emergency plan prepared by the utility with minimal input from local and state officials and to permit operation of facilities on that basis if there is a finding of local or state 'foot-dragging' in the preparation of adequate emergency plans."

Another unsavoury feature of the Diablo controversy concerns Commissioner Kennedy and Commissioner Hendrie — he who lost his chairmanship of the NRC after the publication of the Kemeny Report, only to be reinstated by President Reagan in March 1981. On 20 October 1979 both these Commissioners held a secret meeting with two top PG&E men, Messrs Mielke and Shackleford, who had said they wished to discuss the TMI-related NRC delay in issuing the Diablo licence. The Commissioners, when found out, claimed that they had never mentioned the licence but talked only about "generic scheduling problems and generic procedural difficulties". Nevertheless, both were asked to disqualify themselves from voting on issues relating to the Diablo licence. As the issuing of such a licence requires a vote of the *full* Commission, this could mean an indefinite delay — even supposing all PG&E's other problems had evaporated. A just punishment, the objective observer might think, for their excessive amiability to two Commissioners.

Concerning Hendrie, there is a further complication. Previously he reviewed the Diablo case while a high-ranking staff-member of the NRC, and attended that now-celebrated meeting at which an intervenors' request for a Stop Work order was denied — after the discovery of the Hosgri Fault. The Freedom of Information Act allowed the Centre for Law in the Public Interest to obtain documents revealing the part played by Hendrie in this affair and they promptly based an appeal for Hendrie's disqual-ification upon the NRC's own regulations concerning "separ-ation of functions". Therefore, if Hendrie refused to disqualify himself and insisted upon taking part in future Diablo licensing decisions, a Court challenge would almost certainly succeed.

For years PG&E have been experiencing "seismic" embarrass-ments. On 28 January 1976, a memo from one senior company official, H. R. Perry, to another, B. W. Shackleford, went as follows: "In regard to our discussion on filing and record keep-ing, my current suggestions are: (1) Indoctrinate employees that may be involved into the accessibility of all our files to outsiders and the care that need be exercised. (2) Have all levels of super-vision on the lookout for material that should not be filed. (3) Institute a routine system of minimising material that is kept in files."[8]

On the same date, G. A. Mielke contributed his ideas on this delicate subject in another memo to Mr Shackleford: "You re-quested our suggestions for avoiding the generation of docu-ments which could prove detrimental to the Company at some future time. I believe that the Department head is the key . . . Based on the requirements of his particular area of functional responsibility, he should provide guidelines to his subordinates on what should not be put in writing . . . Continuous monitoring of written material and awareness of potential problem areas should greatly reduce the likelihood of detrimental material getting into the record . . . Minutes of meetings and memoranda to the files should document only actions taken and conclusions reached and not how these were arrived at; and should avoid identifying the views of specific individuals or quoting them directly . . . As a final comment, I would recommend that we shy away from a centralised Company Records Management System, as a means of better monitoring the contents of sensitive documents and possibly expunging them of damaging disclosures. Such systems have, as logical adjuncts, powerful indexing and rapid retrieval capabilities. With such a system in being, it would be relatively easy for potential detractors to identify and retrieve those documents which likely could contain material detrimental to the Company."[9]

There is a nice irony in the fact that these two memos were not expunged, as their writers would surely have wished them to be.

On 28 May 1980 a Los Angeles firm of attorneys, Glassman and Browning, issued the following press release: "Charging PG&E with fraud, conspiracy, and negligence for failure to fulfil the lease terms (concerning safety regulations) for the property where the $1.8 billion Diablo Canyon nuclear plant is located, attorneys for the Marre Land and Cattle Company today filed a $1 million civil damages suit in a San Luis Obispo County Superior Court. The Marre Land and Cattle Company owns the land PG&E leases for the plant site . . . After geologists reported the existence of the Hosgri Fault, PG&E and the federal Atomic Energy Commission (forerunners of the NRC) which issued the construction permits for the reactors, attempted first to cover up the information and later minimise its importance . . . Although PG&E refused to acknowledge to nuclear safety officials the seriousness of offshore faults in mid-1970, the Marre suit contends that publication of the existence of the Hosgri Fault within three miles of the Diablo Canyon nuclear facility, moved PG&E to take internal measures to suppress and cover up information about the fault."

Corruption is endemic within large corporations and we tend, rightly or wrongly, to shrug it off as something beyond our control — like earthquakes. In general, however, corporate duplicity does not have the potential for killing large numbers of people — only adversely affecting their "quality of life", or cheating them, or in certain cases injuring them. But the combination of nuclear fission and Big Biz viciousness is one which we dare not tolerate. A book given to me by Roger, to read on my flight home, documented the consequences of a degree of corruption within America's nuclear industry that in the long term threatens us all.

"The Nugget File" was published in January 1979 by the Union of Concerned Scientists, in Cambridge, Mass. The Introduction explains:

The official optimism about nuclear power plant safety is based on the claim that unprecedented meticulousness is achieved in all aspects of the design, construction and operation of these unique facilities. The confident official pronouncements notwithstanding, there are now a number of doubts ... The controversy over nuclear safety involves two quite distinct types of questions. First, there are *technical questions* that relate to the adequacy of basic nuclear plant design features. No less important are the *institutional questions* that relate to the way in which the people who build and operate nuclear plants carry out their safety responsibilities. The operating records of US nuclear plants provide an important collection of source material that can and should be used to answer some of the technical and institutional questions relating to nuclear safety. This type of information is a vital supplement to the theoretical analyses and laboratory data used to design nuclear plant components, systems and structures: the operating records provide performance data that can be used to judge the adequacy of the industry's nuclear safety efforts. The NRC, however, makes no systematic effort to review the operating plant records, to follow up on problems in individual plants that may affect large numbers of plants, or to reassess safety requirements in a timely fashion in light of reported safety problems. Reports of serious equipment malfunctions are simply buried in the blizzard of trivia that blows back and forth between the operators of each nuclear power station and the NRC. To gain access to useful and important data requires an expensive, time-consuming and frustrating file search. At least, this was the situation until the Nugget File was uncovered by the UCS as a result of a series of Freedom of Information Act requests aimed at compelling the NRC to make a full disclosure of its data and studies on safety problems ... The Nugget File is special, internal file, maintained personally by Dr Stephen H. Hanauer for the last ten years, on serious accidents and safety deficiencies ... As one of the most senior officials in the US nuclear power programme, Dr Hanauer was routinely and promptly informed of safety problems that were detected in operating nuclear plants . . . For his own information he assembled the most significant items into what he labelled "The Nugget File"— a collection of short reports about a wide variety of astonishing safety deficiencies at US nuclear power plants . . . For want of fuses, key nuclear safety equipment is rendered inoperative. Electrical relays fail because they are painted over or welded together or disconnected. Valves in safety equipment are destroyed because switches on their motors are incorrectly adjusted. A 3,000 gallon radioactive waste tank is found in an operating plant to be connected with the plant's drinking-water system. Emergency power sources fail routinely, bizarre equipment failure modes are a common occurence and operator errors are an endemic feature of the US nuclear programme. Simple maintenance operations disrupt established safety precautions, as when valves that affect the water level in the reactor were accidentally shut off during the repair of a leaky faucet in a plant laboratory. Sensitive pieces of safety equipment malfunction because they are frozen or burned or flooded or dirty or corroded or bumped or dropped or overpressurised or unhinged or miscalibrated or miswired; they are also shown, to cite one of Hanauer's marginal comments, to be "guaranteed not to work" because of bad initial designs . . . Dr Hanauer added a marginal comment on one of the Nugget File documents that succinctly expresses the only editorial comment that we would wish to make on the mishaps plaguing the nuclear industry. "Some day," he wrote, "we all will wake up."

"The Nugget File" provides riveting but spine-chilling reading; I hardly noticed the hours passing as I sped across the Atlantic. America's nuclear industry is not funny, yet this book's cumulative effort was to make me giggle — semi-hysterically. My favourite quote: "A diesel generator failed to start. The cause of the malfunction was identified as A LOOSE SCREW." (Editor's capitals.)

4

Further Aspects of Human Frailty

> Concern for man himself and his fate must always form the chief interest of all technical endeavours . . . never forget this in the midst of your diagrams and equations.
>
> ALBERT EINSTEIN

Soon after my return from the US, a Canadian nuclear scientist took me on in the correspondence columns of *Blackwood's Magazine*. He pointed out that in a debate with a writer he was at a disadvantage because "most nuclear scientists are more used to dealing with scientific facts and engineering materials than with human frailties".[1] This was easy to believe: the Kemeny Report had emphasised: "As the evidence accumulated, it became clear that the fundamental problems are people-related problems, and not equipment problems. The most serious 'mindset' is the preoccupation of everyone with the safety of the equipment, resulting in the down-playing of the importance of the human element in nuclear power generation. The NRC and the industry have failed to recognise sufficiently that the human beings who manage and operate the plants constitute an important safety system."[2]

Nuke jargon describes potentially lethal mishaps as "incidents" or "events". These may be caused by equipment failure or they may be "inadvertent criticalities" — that is, serious accidents caused by human error. As "The Nugget File" reveals, human error contributes to most nuke accidents and TMI was but one of many "inadvertent criticalities" which could have had catastrophic results.

On 12 December 1952 human frailty really took over at NRX, one of Canada's first two research reactors on the Chalk River. A technician opened three or four valves in error. Moments later, the supervisor telephoned his assistant instructing him to press buttons 4 and 2 to correct the technician's mistake. That was a second error: he had *meant* to say '4 and 3' but in the stress of the moment didn't . . . Then he and another staff member rushed with a bucket into an area of the plant where water was doing something it didn't oughter. They assumed this to be heavy water — a third error: it was radioactive light water coolant. The entire plant was then evacuated apart from the control-room staff, who remained at their posts wearing gas-masks. The reactor core was almost completely demolished but, if the industry can be believed, safety measures worked so well that no staff member was exposed to excessive radiation during the accident. However, the subsequent clean-up exposed employees to dosages well above the recommended levels.

At Millstone-1, a reactor owned by New England Utilities, an operator failed to insert selected control rods upon load rejection, as designed — this was during start-up testing. In boiling water reactors, if the predetermined sequence is not adhered to in the handling of control rods, a serious accident may occur later if the rod is ejected from the core. The failure of the Select Rod Insert feature in December 1970 was caused by changes made during the start-up test programme. Following a modification of the circuit, nobody had performed a functional test of the system. The utility was fined $15,000 by the NRC and both the erring operator and his supervisor had to be retrained and recertified.

In December 1977, again at the Millstone plant, a door was blown off its hinges and a worker blasted forty feet away. He received burns and lacerations and was contaminated; I made enquiries about his present state of health but received no reply. On that day there were two explosions at the plant, three hours apart.

The Pre-Industrial Age and the Space Age coincided at Brown's Ferry twin reactor on 22 March 1975. Then an electrician used a candle to test for air-leaks and the highly inflammable foam barriers caught fire. The blaze that followed caused so much damage that a meltdown was avoided only at the last moment by jury-rigging the few small pumps still available. The one downwind radiation monitor was not operating but the NRC asserted that there had been no release of radioactivity. They never explained how they knew this: perhaps they were favoured with a Divine Revelation. The total cost of repairing the damage and providing alternative power was at least $150 millions. Thus Brown's Ferry became the world's most expensive industrial accident. But TMI has robbed it of this distinction.

A rarely considered aspect of the human error problem has been discussed by Gregory Minor,[4] who became so uneasy about the US nuke scene that he resigned from General Electric. Just what sort of person is best suited to operate a nuclear plant? When all is going well, the operator's job is stupefyingly tedious. He spends his whole shift watching dials and charts which show little change and make no demands on his initiative or ingenuity. If intellectually well-endowed, he would soon go crazy. Yet his job is crucially important: during crises in LWRs it is he who must make instant decisions. At TMI two average young Americans had to interpret, within moments, over a hundred signals. And Craig Faust's intelligence may be gauged by a remark he made to one of the President's advisers: "The only accident is that the thing leaked out. You could have avoided this whole thing by not saying anything. But because of regulations, it was disclosed."[5]

Walter Patterson, himself a nuclear physicist, refers to "the antiseptic remoteness of the control centre, in which even a major malfunction will manifest itself only at the end of a long sequence of intervening communication links. A worker on an assembly-line may also find the work boring; but at least he will be surrounded by activity. The control room of a base load power station in normal operation is industrially akin to sensory deprivation."[6]

A contributor to *The Economist,* inspired by TMI, also glanced at this problem on 7 April 1979. "Nuclear power station operators (in Britain) are trained in roughly the same sort of way as airline pilots — prolonged practice on simulators, demanding tests, and so on. This analogy is both encouraging (high skills, intensive training) and slightly worrying. Airlines themselves now worry about the lack of crisis-management skills in younger pilots trained exclusively in airliners equipped with a whole range of automatic safety devices ... Should the nuclear industry be looking for smaller, lower-powered, less efficient, but inherently safer designs: ones the average operator can see whole in his mind?"

In the course of our duel, my Canadian adversary asserted that to anti-nukes "the traditional institutions of free enterprise (unless it is very small scale), democratic government, the legal profession and any expert opinion (unless these fully agree with our point of view) have become the enemy".[7] He seems to imagine — perhaps he has lived in a remote part of Canada for a very long time — that free enterprise and the nuclear industry are in some way connected. Not so, however. In February 1977 the British government introduced the Nuclear Industry (Finance) Bill which allocated £1,000 million of public funds for the support of British Nuclear Fuels Ltd. This Bill received the Royal Assent in early April. If that is free enterprise I'm an atom bomb. In no country could a civil nuclear industry have been founded on an ordinary commercial basis; the capital involved is off the commercial scale; the lead-times are too long, the technology is too complicated. Civil nuclear development was made possible only by the weapons programmes of the countries concerned. And we have seen something of the horrific effects of pseudo "free enterprise" in America's nuclear power industry, which at present is being kept afloat only with the aid of massive government-sponsored Runaway Reactor loans to foreign countries.

Many people are unaware of the genesis of the nuclear power industry. In 1951 the AEC made available classified atomic data, and enormous research subsidies, to a group of companies including General Electric and Westinghouse. The latter agreed to run a plutonium-production plant at Hanford for the government's weapons programme, in exchange for a government-provided nuclear-power-development laboratory at Schenectady. It seems such blandishments were necessary because certain utilities initially felt reluctant to tangle with so daunting a technology. Other companies took up the nuclear option only when the AEC implied that if deprived of utility cooperation they themselves might enter the electricity business.

Alexander Cockburn and James Ridgeway have explained: "A major impetus for nuclear power came from the advocates of public power, especially the Tennessee Valley Authority. The TVA had been headed by David Lilienthal, who later became the first Chairman of the AEC. Indeed, the role of the TVA had immense repercussions for the energy industry overall. In its early years, the TVA had concerned itself largely with hydro-electric projects, with the aim of modernising the Tennessee Valley. As the Cold War developed, the TVA found itself having to expand its operations very rapidly for specific military reasons. As a consequence of the Arms Race, larger and larger amounts of enriched uranium were needed for atomic bombs. The process by which uranium is enriched requires vast amounts of electricity. Much of that initial electricity was provided by the TVA, which sent the power to the AEC enrichment plant in Kentucky. The hydroelectric facilities of the TVA were not able to meet the demands of the enrichment plant during the 1950s and in order to step up generating capacity the Authority turned to coal. To this end the TVA sponsored, through a series of unusual (at the time) long-term contracts with coal companies in Appalachia, the re-organisation of the coal industry. The industry was mechanised, small operators were driven out, and the way was paved for the subsequent concentration of the coal industry, which finally fell under the control of the oil industry in the late 1960s. The process then became circular. The TVA created the electricity which enriched the uranium. These supplies of uranium, at first used to make bombs, later became the fuel for the nuclear power plants. Among the most important customers was TVA itself which needed nuclear fuel to generate electricity in its reactors, which was used in part to create yet more fuel. The US is now left to deal with the consequences of that dire bargain."[8]

In Britain, too, the nuclear industry was government-inspired and at first the generating boards cooperated only reluctantly, being appalled at the costs involved and rather resenting their role as lackeys to the military Chiefs of Staff. The Harwell and Risley reactors were designed in the late 1940s to produce weapon-plutonium. Then a uranium factory went up at Spring-fields in Lancashire and an enrichment plant at Capenhurst in Cheshire. Windscale, in Cumbria, was favoured with a prototype AGR, two plutonium-producing reactors and a plutonium-separating plant. Both plutonium-producing reactors died young, after the 1957 fire. The full Fleck Report on that disaster (after Sir Alexander Fleck, who chaired the enquiry) has never been published. But Walter Patterson comments: "It is clear that the design of the reactor, especially of its instrumentation, was severely flawed in some essentials; and plant administration also left something to be desired."[9] It still does, as we all know. On 26 February 1981 BNFL was fined £500 with £133.86 costs, after pleading guilty to a breach of safety regulations at Windscale. Commenting on the case, Mr Leslie Clarke, of the Health and Safety Executive, said: "That an enterprise such as BNFL should fail on so many counts in the organisation of radiological protection is lamentable. Arrangements had fallen significantly below minimum standards." Windscale always has been a hideous blot on the British nuclear industry's copy-book.

In October 1956 the "switching on" of Calder Hall by the Queen caused tremendous excitement. This reactor was acclaimed as "the world's first nuclear power station" though in fact it was just another nuclear-weapons installation which had been linked to the national grid. Both Calder Hall and Chapelcross in Scotland, which came on stream at about the same time, can still be used to produce weapons-grade plutonium.

Walter Patterson has explained: "The design work, materials research and prototype development not only of reactors themselves but also of their ancillaries were carried out under the auspices of the government nuclear agency . . . The Atomic Energy Act (1954) created the UKAEA, which was given responsibility for all research and development of applied nuclear energy, military and civil. The AEA was, and still is, in some respects a unique arm of the government. It was funded by a separate Vote of Parliament, called for many years the Atomic Energy Vote. In February 1954 the first civil Estimate under this heading was presented, for the sum of £53,675,000. Over the years the annual Vote has slowly increased. It is now called "Industrial Innovation: Nuclear Energy". In 1975–6 the AEA's Estimates anticipated gross cash expenditures of £151,871,000 . . . The nuclear energy Vote places this form of energy on a footing unlike that of any other in the British economy."[10]

Pro-nukes' references to "democratic government" suggest that they are unaware of the fact that an expanding nuclear industry must inevitably undermine civil liberties. Several US government-sponsored studies have stressed the danger should international terrorism "go nuclear" and security experts have recommended wire-tapping, increased surveillance of dissident individuals and the infiltration of anti-nuke organisations to detect possible plots against nuclear stations or military bases. Many citizen groups such as The Sierra Club, Another Mother for Peace, Environmental Action, FOE and the UCS are now kept under constant surveillance.

Continental Airlines pilot Robert Pomeroy, a member of the Citizens' Association for Sound Energy which some years ago opposed the building of a reactor near Dallas, was investigated by the Texas State Police on the suggestion of the FBI. Continental Airlines were "warned" about Pomeroy's activities but instead of dismissing him they told him that he was being investigated, to the discomfiture of the police — if "discomfiture" does not imply an improbably mild reaction on the part of Texas cops.

A few years after the Pomeroy case the reporting staff of the *Nashville Tennessean* was infiltrated by an FBI informer named Jacqueline Srouji; her job was to spy on an editor and reporter who had written anti-nuke articles. Such actions are suggested by the Federal Government and the utilities, whose propaganda machine incorporates the Atomic Industrial Forum and the public relations firm of Charles B. Yulish Associates in New York City. Thus nuclear power directly threatens personal freedom. And eventually a vicious circle could form, if those who most resent undemocratic restraints were driven to attempt to reform society through political violence. The nuclear industry has created a world in which nuclear terrorism and sabotage are an ever-present danger, yet it is the "dissidents" who are being treated as criminals. And if the pro-nuke Establishment continues to stand on their necks, some of them may turn to anti-nuke crimes.

In Britain, as Flowers (para. 332) noted: "What is most to be feared is an insidious growth in surveillance in response to a growing threat as the amount of plutonium in existence, and familiarity with its properties, increases; and the possibility that a single serious incident in the future might bring a realisation of the need to increase security measures and surveillance to a degree that would be regarded as wholly unacceptable, but which could not then be avoided because of the extent of our dependence on plutonium for energy supplies. The unquantifiable effects of the security measures that might become necessary in the plutonium economy of the future should be a major consideration in any decision concerning a substantial increase in the nuclear power programme."

At present, plutonium fuel is regularly moved from the BNFL fuel fabrication plant at Windscale to the AEA experimental FBR at Dounreay. The strict security embargo on information about these movements must naturally apply to Parliament as well as to the public. And this is of some significance. The AEA has its own special constabulary to guard both installations and SNM transport — a force empowered to carry arms at all times, to engage in "hot pursuit" of those stealing or suspected of trying to steal SNM, and to arrest on suspicion. The passage through Parliament of the Atomic Energy Authority (Special Constables) Act 1976 indicates that plutonium security must now take precedence over those democratic controls which the British have enjoyed for longer than most people. Before and during the Bill's passage, a number of peers and MPs urged that the establishment of an independent AEA armed police force should somehow be avoided. Several peers and MPs wished to see the army taking over the guarding of installations and transport but the government insisted upon the creation of the Special Constabulary — whose standing orders will not be published and whose accountability to Parliament is tenuous. Many of Her Majesty's subjects, not only anti-nukes, are alarmed by this creation of a private armed force whose composition, organisation and conduct are not open to Parliamentary or public scrutiny. Alan Beith MP thought the Bill could "be regarded as a very dangerous threat to civil liberties".[11] And Lord Mansfield argued that it provided altogether inadequate Parliamentary controls over a situation "wholly new in our history", involving "a quite new departure in our constitutional thought on carrying arms".[12]

According to the "Report of Lord Radcliffe's Committee, 1962" on Security Procedures in the Public Service, the AEA relies on the "Security Service to supply information and advice on the nature and scale of the various threats which exist to the security of the State and to advise on measures of defence against them" (para. 15). The Security Service means MI 5, which is unrecognised in law and "operates independently under its own

Director-General" (Lord Denning's Report, 1963, para. 273). The Director-General is responsible only to the Home Secretary for MI 5's "efficient and proper working". Other cabinet ministers and MPs have to keep off the grass and individuals with a grievance against MI 5 can appeal to neither court nor Ombudsman. The fact that MI 5 is in charge of nuclear security means that the public cannot find out what diminution of their civil liberties may ultimately be involved in the creation and transportation of plutonium.

Britain's Official Secrets Act is another menace, in the nuclear context, as it insulates the industry from public criticism. Careless operators, errors in design or construction, missing plutonium, neglected safety precautions and muddled thinking within the nuclear establishment are shielded from scrutiny by this Act. It also deprives Parliament of the sort of independent advice on nuclear matters that is available to the US government (not that it does them much good). In 1972 the Vintner Report, a departmental study on thermal nuclear reactors, was withheld from the Commons Select Committee on Science and Technology by the Department of Industry. "Considerations of commercial confidentiality" were given as the excuse. In the US, however, the most constructive critics of the nuclear industry are former AEC employees who have "inside information" and use it publicly. If their British counterparts did likewise, they would be inviting arrest and imprisonment.

On one extremely important safety issue some American authorities show much more common-sense than their British counterparts. In New York there is a total ban on the transport of spent fuel or radioactive waste through the city. This was brought about by Dr Leonard Solon, Director of New York's Bureau of Radiation Control, who sees the consequences of a major accident as entirely unacceptable. "With a massive release we're talking abou. the potential of thousands of long-term malignancies or cancer deaths and hundreds of prompt deaths . . . The concern is not with the flasks or casks that survive testing. You're concerned about the five out of a hundred, or two out of a hundred, or one out of a hundred in which there is some kind of engineering defect or imperfection in which some quality assurance procedure had failed."[13]

Dr Solon believes that London should impose a similar ban on the spent fuel that is carried by British Rail to Windscale from Sizewell, Bradwell and Dungeness. Between 1962 and 1980, at least 9,000 tons of spent fuel were transported through London in about 4,500 consignments. (That is the CEGB figure; BNFL say 12,000 tons and the AEA say 15,000 tons.)

How likely is an accident? All attempts to answer that question accurately are frustrated by the British nuclear industry's obsessional furtiveness. In the case of spent fuel transport, one can only deduce the industry has a lot to hide, particularly in relation to the design and testing of flasks.

In January 1978, when Tony Benn was Secretary of State for Energy, he submitted a detailed questionnaire to the various atomic energy establishments and received alarmingly evasive answers. His first question concerned Safety Related Incidents (SRIs): "What documents contain a list of all safety related incidents at Windscale, and to flasks containing irradiated fuel, since 1950?"

The reply: "There is no document listing SRIs involving flasks containing irradiated fuel. Such incidents would have been reported by operators of licensed nuclear sites under the Nuclear Installations (Dangerous Occurrences) Regulations 1965, or in the case of the UKAEA, under equivalent arrangements. *No such incidents have been reported*." (My italics.)[14]

Two years later, Mr Kenneth Clark, Permanent Under-Secretary, Department of Transport, stated in Parliament that, since 1976, eight instances of minor contamination above the permitted levels (involving spent fuel flasks) had been reported to the Health and Safety Executive.[15] So that was *eight* SRIs in *three* years . . . How many, then, have in fact been reported since 1950? And why does the industry make such puerile attempts to cover its radioactive tracks?

Mr Benn's sixteenth question asked: "BNFL document no. WA 38A presents "laundered" summaries of 45 of the 177 accidents or SRIs listed, or rather mentioned, in question 10. Will you get the full report for us to review?"

The reply: "The summaries provided in WA 38A present reasonable retrospective accounts of past incidents . . . It would not be practicable to provide full reports because they contain information of a proprietary or confidential kind about plant or personnel." The Ecology Party has commented: "This is the industry's favourite excuse for refusing to release information about its work, and it raises a fundamental issue: accountability. If the nuclear industry can withhold information, on one pretext or another, from the Secretary of State for Energy, an elected Cabinet Minister ostensibly representing the people of this country, then to whom does it consider itself accountable? Regrettably, the answer to that appears to be 'only unto itself'."[16]

On 3 July 1979 an article in the *Guardian*[17] disclosed that faulty welding had been detected in twenty-two stainless-steel bottles used as containers for irradiated fuel elements. The CEGB at once disowned those bottles, which apparently were being used by BNFL to transport *foreign* spent fuel to Windscale — a distinction that might seem irrelevant should such defective containers ever happen to irradiate a section of the British public.

Potentially catastrophic spent fuel moves through London as freely as loads of old rope — and sometimes *doesn't* move, but lies around unsupervised for hours or even days, tempting Providence. On 1 November 1979, soon after midnight, three men armed with a rocket-launcher, and making no attempt to conceal it, sauntered up to a flask standing at Platform 9 on Stratford station. They aimed their weapon at the flask and nobody was interested. They might have been Provos; luckily for London they were members of the Freedom of Information Group, whose rocket-launcher was a theatrical prop. Subsequently the British Railways Board tried to cover its blushes by declaring: "It is not the job of BR staff to apprehend people carrying rocket-launchers . . . The men had a perfect right to be on the platform provided they were in possession of a valid ticket."[18]

All this is the more disquieting because Britain's flask designers seem unaware of — or indifferent to — the power of modern weapons. Peter Faulkner has explained: "The extremely rugged casks used to transport spent-fuel elements provide a protective layer consisting of about two inches of steel and eight inches of lead. However, they are penetrable by one or more shaped explosive charges developed to pierce military armour. A hemispheric charge made of a few pounds of high explosive lined with any dense metal will penetrate two feet of solid steel."[19]

The Ecology Party has noted: "There is a belief, widely held in official circles, that terrorists will deem it more profitable to attack some other kind of hazardous cargo which is more commonplace, less well encased and more dramatic in its immediate effects. This line of thought . . . fails to take into account the emotive value of a radiation scare — about the only event more likely to induce mass hysteria would be an invasion from outer space."[20]

If reactor fuel is deprived of its coolant, the *decay heat* can cause overheating; and the same process could occur within a fuel flask, though no flask contains enough active fuel to sustain a chain-reaction. This overheating possibility was one of many reasons for Flowers' (paras 420, 421) primly expressed alarm: "We were surprised to learn that the tests are conducted only on models, and since the containers travel on ordinary freight trains which may be expected to travel at speeds up to twice that assumed in the tests (with kinetic energy four times that assumed), we were not wholly reassured."

Spent fuel is allowed to share freight trains with petroleum tanks and other inflammable liquids, though BR's staff manual states that these should not travel *next to* a flask wagon. (How cautious can you get!) During the 1968–77 period, there were ninety collisions between passenger trains and freight trains or light locomotives. And there were 184 collisions between freight trains and light locomotives or other moving vehicles. Twenty-two collisions were the subjects of enquiries, as "serious accidents", and at least eight happened on flask-carrying lines.[21]

A new transport risk has now been introduced by the AEA — the shipping of plutonium nitrate from Scrabster, near Dounreay, to Workington, near Windscale. It has already been proved that a nuclear cargo can be "diverted" between ports and the Scrabster-Workington route is a lonely one. Will the 2,355-ton *Kingsnorth Fisher* have a navy escort? Or will a naval equivalent to the AEA Special Constabulary be established? Here hijacking is the main problem: but safety, too, is a valid concern. At sea, a radiation-releasing accident is more likely than on land. These cargoes will require the strictest supervision; if not kept below a certain temperature, a large mass of highly radioactive plutonium nitrate could very quickly go on one of those celebrated "excursions" (i.e., blow up). It is improbable that the *Kingsnorth*

Fisher will ever sink, though the Minch is frequently stormy, but if it *did* nobody can guarantee that its cargo could be retrieved intact. Nor is any clean-up possible in a contaminated sea.

Pro-nukes boast that for eighteen years spent fuel has been transported around Britain without any serious accident: from which happy circumstance they deduce that all is — and must continue to be — well. But caprolactam plant seemed safe until Flixborough; and dioxin until Seveso; and system building until Ronan Point. Whenever the nuclear industry congratulates itself on its safety record, one can only say, "Wait for it!"

On one issue — exploratory drilling in search of high-level waste burial sites — I am rather out-of-step with my British anti-nuke friends. This quest for safe "graveyards" is an ethical hot potato. At a certain point one persuasive gentleman from Harwell, who for years past has been working on the vitrification of wastes, almost convinced me that he had the answer. And perhaps he has. But Flowers (para. 181) emphasised the snag here, in a much-quoted paragraph: "It would be morally wrong to commit future generations to the consequences of fission power . . . unless it has been demonstrated beyond reasonable doubt that at least one method exists for the safe isolation of wastes for the indefinite future."

How *can* such a thing be demonstrated "beyond reasonable doubt"? How much do nuclear scientists *know* about their idol's potentialities? Can they guarantee that such-and-such a cere-mony — like fusing actinides with glass, enclosing the cylinders in nickel-chromium steel which must then be enclosed in copper, lead or titanium, and burying them thousands of feet under the earth or the sea-bed — can they guarantee that this astronomically expensive ceremony will indeed sufficiently propitiate the God of the Underworld to prevent his harming humanity "for the indef-inite future"?

In July 1979 the British government named fifteen sites where the AEA planned to carry out test-bore drilling in search of safe graveyards. Explaining the problem, Richard Woodmank, the Press Association science correspondent, wrote: "Sites would have to be so deep that they would not be affected by glaciation if there was a new ice-age. Yet they would need to be high enough above sea-level not to be flooded by the sea if the polar ice-caps melted . . . The AEA has always insisted the drilling is only to help geological research but planning applications have met with stiff resistance from conservation groups and local councils."[22]

It is of course splendid that the British are becoming less and less willing to tolerate that obtuse immorality which allows the creation of plutonium. But lethal quantities of high-level waste *already exist* and for this graveyards *must be found*. So it is arguable that to compound the difficulties of finding the safest spots, by staging demonstrations or legal wrangles, is inconsistent with a concern for humanity's welfare. In Britain, Windscale at present stores all this waste and the place is a menace — perhaps more so than we yet realise. Therefore the sooner a suitable graveyard can be found the healthier our world will be. This may indeed mean passing the nuclear buck to future generations: but keeping that buck above ground, in a place like Windscale, is helping neither our descendants nor ourselves. Or so it seems to me.

However, a number of the most responsible British anti-nukes insist that the industry must not be left free to pronounce an unsuitable site suitable. Graveyards are its big worry at present: without them it will be increasingly threatened by public anxiety about growing dumps of unmanageable actinides capable of doing incalculable damage. Seen from this angle, there is a strong case for organising the maximum opposition to test-drilling — especially as "experts" are still disagreeing about the fundamentals of waste-disposal.

In a letter to *The Times* on 12 November 1980, D. C. Leslie, Professor of Nuclear Engineering at Queen Mary College, assured us that: "There is little doubt that the final disposal should be in some geological formation. Therefore, the question is not so much whether these cylinders should be buried, but when and where . . . The research programmes of the EEC countries into the disposal of nuclear waste are carefully coordinated, with different countries looking into disposal in different types of rock strata, and they are going well. The UK and France are investigating disposal into crystalline rocks such as granite."

Then, on 19 February 1981, the Press Association reported that a British National Radiological Protection Board study assumed that ground water would leak into nuclear dumps after about 1,000 years. Dr Marion Hill, a BNRPB scientist, had therefore advised a coastal site, arguing that from the sea-bed "radioactive elements were unlikely to end up in drinking water". Instead, sea water would so dilute them that the environmental damage would be "several orders of magnitude less than with an inland site".

All of which underlines the point that nothing can be known *with certainty* about the long-term behaviour and effects of nuclear wastes, despite the vast amounts of money expended on their study. Pro-nukes are extremely reluctant to divulge how much has already been spent on this branch of research and development (r & d); and they are even more reluctant to discuss in detail the economics of their "safe" waste-disposal programmes. According to Dr L. E. J. Roberts, Director of the Atomic Energy Research Establishment at Harwell: "Early estimates of the cost of vitrification were that the capital costs would be less than 1 per cent of the value of the electricity generated by the programme giving rise to the wastes and there is no reason to depart from this view."[23] But where are the figures on which these early estimates were based? And how can Dr Roberts be so confident that they remain — if they ever were — valid? His own description of waste disposal r & d, at Harwell and elsewhere, strongly suggests an astronomical bill for the taxpayers.

Britain's nuclear problems are as nothing compared to those of France. Monsieur Puiseux, an economist working for Electricité de France, regards a nuclear-dependent state as "a society full of police".[24] A French Parliamentary Commission has expressed concern lest nuclear decisions might have been taken under "pressure" from the American industry. And in 1976 a French ecumenical church conference found that "the actual programme for nuclear energy production is based on the plan put forward by the Westinghouse Corporation, and decisions affecting our country have never been the subject of parliamentary or public consultation".[25]

In 1969 France was the scene of a remarkable coincidence. On 16 October, when the country's last graphite-moderated gas reactor was inaugurated at St Laurent-les-Eaux, Monsieur Marcel Boiteux, the Director-General of Electricité de France, made a curious speech in which he emphasised not the admirable qualities of the reactor he was inaugurating but the superior merits of LWRs. (Gas reactors are unpopular with the multinationals, who have put their collective shirt on LWRs.) The experts in the audience were puzzled, finding it hard to see why a native French series of designs should be abandoned in favour of a foreign series. No disinterested nuclear scientist had ever faulted France's home-grown reactors — until 17 October 1969. Then, on the day after its inauguration, the St Laurent-les-Eaux plant was put out of action by the *repeated* mistakes of an experienced operator. This man continued to make false moves until several of the new uranium rods had fused, severely contaminating the container. For a year 300 people worked on the clean-up and one received a 5,400 millirem dose — well above the tolerated limit, even for employees.

From then on the LWR merchants were winning. Their final victory came on 4 March 1974 when the Prime Minister, Pierre Messmer, decided that American reactors should replace those native products which had been proved so very unreliable at St Laurent-les-Eaux. Another coincidence may be observed here. President Pompidou, who had many and serious doubts about the ethics of nuclear power, was dying. As soon as he lost control of the government, Monsieur Poujade and Monsieur d'Iribarne, the ministers most opposed to the nuclear programme of Electricité de France, were replaced. Only forty-eight hours before the pro-American decision was made, Monsieur Poujade — an exceptionally well-informed conservationist — was succeeded by Monsieur Peyrefitte, who had neither the knowledge nor the will to argue against the pro-Westinghouse forces.

Can Monsieur Messmer have been aware of all the implications of his decision? A 1975 report on nuclear alternatives, prepared at the Grenoble University Institute of Energy, pointed out: "So far as the construction of LWRs is concerned, the methods of construction are laid down by the Americans. The French constructors have no details about the way in which these reactors have been developed, and what is more, the methods relate to types of reactors which will not enter into service in the US until after the French power stations now under construction have been inaugurated."[26]

If Electricité de France achieves its ambition, "All electric, all nuclear by AD 2000", France may become even more dependent than she now is on foreign fuel sources: yet "fuel independence" is the main plank in the propaganda platform of Electricité de France. Worse still, she will become alarmingly dependent on foreign technology. All nuclear safety matters and plant repairs are covered by a treaty of commercial secrecy with the manufacturers and only they have the power to decide when production should be stopped for safety reasons. Yet, despite this power, the manufacturers cannot be held responsible for any accident, or additional cost, even if these are indisputably the results of flawed equipment or ill-trained operators. On commercial secrecy, the Grenoble Institute of Energy commented: "When energy options are chosen by official fiat or decree, by ill-informed politicians, it is logical that further information should not be given to the public ... In nuclear matters, the monopoly of information belongs to the public powers which carefully select that which may be said and that which may not be said. At the regional level, the consultations with which local representatives are supposedly honoured are devoid of sense: the information is carefully filtered and those who do not represent the official position are often not called upon to speak."[27]

In fact the situation in France is even more serious than this report indicates; those who try to analyse the safety and economic aspects of nuclear power, or who ask why Electricité de France began to build power stations before receiving planning permission, may find themselves with a dossier compiled by an official of French Military Intelligence. In 1976, a Study and Research group, appointed by one of the French trade unions for the electricity industry, declared: "It is certain that the nuclear authorities deliberately hide the risks from the public. And research into safety and the environmental consequences are derisory, compared with expenditure on publicity."[28]

M. Jean Servant, a 55-year-old official with overall responsibility for safety in French nuclear installations, resigned towards the end of 1980. His letter of resignation was dated May 1980 but published only on 15 December 1980, in the weekly magazine *Le Point*. According to Anne Sington: "The resignation has unleashed a wave of disquiet concerning the possibility that dangers inherent in the nuclear option are being concealed from the public. Appointed in 1975 to head a committee for nuclear safety, M. Servant became steadily more disillusioned by the lack of cooperation from French government bodies. The Health Minister refused to cooperate because he was opposed to the Industry Minister's monopoly of nuclear safety. The Ministry of the Interior showed itself as chiefly bent on avoiding involvement in 'problems that are new and complex and a little disquieting'. The Industry Ministry started out actively collaborating but 'gradually withdrew both cooperation and information until it arrived at its barely veiled hostility of today'. Harassed and impeded at every turn, M. Servant succeeded in holding only one meeting of the Nuclear Safety Committee — and that was with restricted attendance. 'I never asked for the job,' he said, 'but having been given it I find it hard to put up with the idea of not doing it well and thoroughly.' "[29]

On 4 October 1979, the *Financial Times Service* reported that "Electricité de France has decided to start bringing into operation two new nuclear plants despite continuing trade union protest about faulty metal components . . . The Industry Ministry had earlier given the go-ahead for the projects, saying it had undertaken a particularly detailed study of the safety factors in the two power stations during the last few months. While admitting that there were cracks in certain key reactor components, the ministry's safety inspectorate says they present 'no immediate risk' . . . Loading of the first reactor with enriched uranium was due to start yesterday at the Gravelines plant and later this week at Tricastin. In spite of the authorities' statements, the two main trade unions in the industry . . . insist that they have not received enough assurance on safety . . . Inspections have revealed the faults in components still in the factory, in stations under construction and in reactors ready to go into operation. But it is also thought that a number of stations already in use may have similar cracks which cannot be detected. Machines are being developed by the authorities to inspect the cracks and undertake repair work by remote control in the radioactive areas. These robots should be ready within the next two years. Meanwhile, different manufacturing methods are said to be eradicating the faults in the components. Those parts which have already been produced are to be

repaired where necessary by the electricity board and Framatome, the engineering company which makes the pressurised
water reactors under licence from Westinghouse in the US."

Three weeks later, Monsieur Shoja Etemad, a French nuclear
engineer, wrote about those cracked reactors in the *Guardian*.
For some years Monsieur Etemad had been a system designer
with Framatome. He explained: "Trade Union federations have
forced the French Government to halt the fuelling of the nuclear
reactors, Gravelines I and Tricastin I. This is because major
components in the system . . . have been found to have deep
fissures. I was one of the engineering team which originally
assessed the problem. We found that the cracks propagated most
rapidly under the stresses imposed by normal working conditions. The cracks are irreparable by any technology known at
present. In normal engineering practice the plates should be
replaced. But the material cost, and the costs in loss of time, would
be high. The French authorities have said that the plates are safe
for the time being and that, in five or six years, robots will be able
to go in and carry out repairs. No such technology exists . . . The
French safety authorities did not share our concern and gave the
go-ahead to use them in spite of the fault and without having
found a way of repairing them. These cracks are hazardous in
themselves, but most important of all, they open up the possibility
of another independent failure during any sudden change of
conditions in the system as a result of an accident." Monsieur
Etemad argues that TMI has destroyed the scientific basis for
nuclear system design, though the international safety authorities
have not even recognised the technical issues; and he describes in
detail the collapse of the scientific methods on which safety calculations are based.

To whom can we look for objective judgements about the safety
or otherwise of individual nuclear plants? Those American
experts who have resigned from the industry for reasons of
conscience are naturally ignored — when they are not being
vilified — by the corporations from which they have defected. So
their perceptions can never be placed directly in the decision
scales to counterbalance the pro-nuke experts. However, by
speaking out they have helped to stop nuke expansion in America
and the industry bosses are showing signs of panic.

On the eve of TMI, the *Public Relations Journal* advised those bosses: "Instead of trying to arouse the public *for* nuclear power, we should change course and try to arouse the public *against* the anti-nuclear groups and what they advocate. Give the people something unattractive to be *against* instead of something beneficial to be *for* . . . We need to go on the offensive, to become activists ourselves. It's time to call a spade a dirty rusty shovel . . . So forget the facts once in a while. Counter the activists not with facts but with closed factory gates, empty schools, cold and dark homes and sad children."[30]

This advice was not needed by Westinghouse's nuclear division, which in 1976 produced *One Liners*, a handbook of "concise responses to many of the questions about nuclear power". So concise are these responses that the whole truth can never find room to appear. For instance: *Q.* What is our experience with handling plutonium? *A.* Plutonium has been handled for more than thirty-four years. No deaths or adverse health effects have resulted to any workers or members of the public. *Q.* How did Karen Silkwood, the technician at a Kerr McGee nuclear fuel plant, really die? *A.* The Oklahoma State Highway Patrol decided (*sic*) that Miss Silkwood fell asleep while driving and ran off the road. She had been taking drugs and drinking just before the accident.

When Karen Silkwood's car left the road "under mysterious circumstances" she was on her way to hand a dossier of documents, proving criminal carelessness at the Kerr McGee plant, to two *New York Times* reporters. The dossier was not in the wrecked car when the police looked for it. A short time previously, the food in Karen's refrigerator had been massively contaminated with plutonium. Her three small children have since been awarded $10.5 million against Kerr McGee. The next time you come across a sliver of Westinghouse propaganda, remember that story.

As the anti-nuke movement becomes more widespread and better-informed, pro-nuke tactics become more ruthless. Not long after TMI, the Edison Electric Institute of America inserted a full-page advertisement in the *Atlantic Monthly*. It read, in part: "Unless we try to rely entirely on coal production for electricity generation, or to shut down the economy until forms like solar power prove affordable, nuclear power will become increasingly important to the near-term well-being of the country." This is typical of the alarmist propaganda technique advised by the *Public Relations Journal*.

Next we get: "The electric utilities industry agrees with the many expert opinions that ultimate disposal of radioactive wastes presents no insurmountable problem. Several acceptable methods are available. But failure of the Federal government to implement available nuclear waste disposal technology is being mistakenly seen as an indication that the nuclear waste issue cannot be resolved." This prompts the question, "Acceptable to whom?" The fact that a basically pro-nuke Federal government is hesitating to "implement available nuclear waste disposal technology" strongly suggests that this is acceptable only to the industry itself.

The advertisement continues: "Recently, we urged the Administration to take advantage of extensive, existing technical and scientific knowledge and to implement a program on a rigid schedule to provide a spent-fuel storage facility and a waste repository at the earliest practical time. These steps are necessary to assure the continued operation of nuclear power plants, to minimize the uncertainty that has been slowing down commitments for future nuclear plants in this country, and to separate the waste disposal issue from the licensing of new plants." The monstrous suggestion that the waste disposal issue *should* be separated from the licensing of new plants made me wonder for a moment if there had been a misprint.

Throughout America there is an increasing awareness of the implications of sanctioning new plants while the waste disposal problem remains insoluble. In a Supplementary View to the Kemeny Report, Commissioner Peterson wrote: "The disposal of nuclear wastes constitutes, over the long run, the most hazardous aspect of the nuclear power industry. While the industry waits for the government to finish its decades-long effort to determine how to safely dispose of long-lived wastes, such as plutonium, cesium and strontium, each nuclear power plant continues to store its growing amount of spent fuel containing these wastes in a pool of water immediately adjacent to the containment building. I recommend that a serious study be undertaken of how such storage may exacerbate the threat from accidents or sabotage . . ."

On 29 October 1979 the *Washington Post*'s leader-writer noted that: "The states are playing nuclear Old Maid. They are fobbing off nuclear waste on each other . . . The burial sites in Washington and Nevada have been closed for investigation and correction of unsafe handling practices, and both should re-open. But the South Carolina site is closed indefinitely because Governor Richard Riley doesn't want South Carolina to become the nation's nuclear dump. Not surprisingly, a lot of other governors feel the same about their states." It is evident from this leader that even pro-nukes are fretting about waste disposal, as well they might. The industry reckons that by AD 2000 America will have created 152 million gallons of high-level waste. Now some 84 million gallons have to be dealt with — at a cost of billions of dollars, though the methods used are temporary expedients. Nuclear industrialists, as the EEIA advertisement reveals, wish America's tax-payers to foot the waste-disposal bill. And of course Britain's tax-payers will have no choice but to foot that bill, since the British nuclear industry is run by the government.

Given so many powerful vested interests in the foreground, it is hard for the ordinary citizen to judge how *real* the world's energy crisis is. According to the US Environmental Action Foundation, at least half the stock of ninety-nine major investor-owned utilities was held by their ten largest stockholders, dominated by banks. "A 1973 report by the Senate Governmental Operations Committee, entitled *Disclosure of Corporate Ownership*, revealed that the financial institutions which control a utility may also control some of its large customers. The same New York banks which dominate the power industry also control its chief suppliers of generating equipment, General Electric (GE) and Westinghouse (who also make electrical appliances). These control oil companies too — which in turn control large portions of our energy resources. A given utility may be striving to maximise profits for itself, Chase Manhattan Bank, General Electric and Atlantic Richfield all at the same time."

Westinghouse and GE have been cornerstones, for over eighty years, of the financial empire of J. P. Morgan & Co. This gigantic banking group, now with the Morgan Guaranty Trust at its hub, also controls AT&T, US Steel, several railways, many electricity-generating companies and a substantial percentage of the uranium needed by nuclear power stations. Westinghouse and GE sell 75 per cent of American-produced reactors. Because so much capital is involved in the construction of nuke plants, these and a few other giants — like B&W, who designed the T M I reactor — profit vastly from the development of nuclear power. At present a 1,000 MW unit costs almost $1.5 billion and early in 1979 *Business Week* declared that "the most hopeful" government and industry forecasts for nuclear power growth in the '80s called for 180 new plant orders — worth about $270 billion at 1979 prices.

While the oil companies proclaim the need for "US energy independence", they are deliberately reducing US production and profiting indecently through importing oil from OPEC countries. Working with a small group of European oil trusts, they control the distribution and sale of OPEC oil and can skilfully create oil shortage panics. By manipulating world supplies during the last months of 1978, the five biggest US oil companies made the following profit increases: Exxon — 48 per cent; Gulf — 42 per cent; Mobil — 10 per cent; Standard Oil of California — 33 per cent; Texaco — 72 per cent. President Carter described these profits as "unfair to the American people". But the oil companies are so beloved by so many politicians that nobody can do anything to control them. Now they are buying up more coal companies and uranium mines, by way of tightening their grip on energy sources. There are huge coal reserves in the US but the energy bosses are reluctant to invest adequately in improved railways and in the pollution control equipment needed for the safe conversion of coal into energy.

So — how real is the energy crisis? According to the UCS: "The seventy-two operating nuclear plants in this country (the US) presently account for about 10 per cent of total electricity generating capacity and provide roughly 13 per cent of all electrical consumption. Even so, were all seventy-two plants to be brought off line simultaneously in order to correct safety deficiencies, there would still remain more than enough generating capacity to take up the slack, for the US is currently endowed with a substantial excess of electrical generating capacity."[31]

Having described exactly how nuclear power could be phased out — a point that worries many anti-nukes — this writer concludes: "In the event of permanent plant shutdowns, it remains to be determined whether financial responsibility for defraying the cost of "scrapped" investments will be shouldered by the utility stockholders — who reap the benefit of profits — or by the customers themselves."[32] Considering the scale of this invest-ment, it is hard to believe, even in one's most optimistic moods, that it will be voluntarily scrapped *before* the nuke idols have received their human sacrifice. If it were, however, it is not difficult to foresee who would have to carry the can. It wouldn't be the Chase Manhattan Bank, Atlantic Richfield or Westing-house.

Early in 1979 Dr Vince Taylor presented a report, "Energy: The Easy Path", to the US Arms Control and Disarmament Agency (ACDA). He stressed that future energy shortages could be avoided by "simple measures, such as improving the design of new houses and reducing heat losses from existing ones. Major reductions in energy consumption could be achieved without depriving anyone of desired energy services. This is true not only for the US but for other countries thought to be much more efficient users of energy. Therefore neither the limits to oil nor the possibility that nuclear power may be judged unacceptably dangerous need be cause for serious concern. Detailed analyses show that productivity improvements alone could extend the lifetime of conventional fuels sufficiently to permit them to con-tinue as the dominant sources of world energy until well beyond 2025 — providing time for an unhurried, gradual transition to renewable sources." This view supports the Ford Foundation's 1974 *Energy Policy Project*, which revealed that the US has 200–400 billion barrels of untapped recoverable onshore and offshore oil, in addition to what has been already precisely measured. The upper figure exceeds all known reserves in the Middle East. The Foundation's natural gas estimate was even more impressive: 1,000 to 2,000 trillion cubic feet, of which only one or two per cent is now being used.

Dr Taylor is both a physicist and an economist and has worked for RAND and Pan Heuristics, a Los Angeles consulting firm which sponsored several of his influential studies for ACDA. But these employers were upset by his latest effort. He explains: "Apparently, because 'The Easy Path' argues against present US policy — which emphasises the importance of nuclear power — both ACDA and Pan Heuristics disclaim responsibility though they paid for most of it."[33]

In the UK, Gerald Leach and his team — at the end of a two-and-a-half year study sponsored by the Ford Foundation — found that Britain now wastes so much energy that the national wealth could be trebled during the next fifty years without increasing electricity consumption. The Department of Energy has never made any coherent comment on this finding, possibly because so much money has already been poured into Britain's nuke programme that to take the Leach Plan seriously would be to admit to governmental misspending on an unprecedented scale.

Energy conservation would also lower the West's unemployment figures. Around Carnsore, Co. Wexford, where a crazed Irish government was planning to build Ireland's first nuclear power station before the recession mercifully intervened, many locals see the project as an almost magical solution to unemployment. Such a plant might employ some 2,000 workers for at least five years (perhaps fifteen, allowing for Ireland's strike rate), but the majority would have to be highly skilled outsiders. When I remarked on this the locals replied, "Well, anyway they'll have great pay and they'll be spending it all around Wexford" — exactly the argument used around Harrisburg when anti-nukes were campaigning against the building of TMI. And of course extra cash does flow through such "privileged" areas. But there are snags; many nuclear power stations are built in comparatively remote places where the resources of the nearest small towns are unequal to the demands made by an influx of workers and their families. The results are housing shortages, overcrowded schools, the overtaxing of water and sewage supplies and the unsettling of the local community. Then one day most of those skilled workers go home and are replaced by a few hundred white-collar workers who will operate the plant for its lifetime of perhaps thirty years.

In contrast, the US Senate Commerce Committee has estimated that 400,000 jobs would be generated by a $1.6 billion investment in interest subsidies and loan guarantees for insulating buildings. This is two hundred times as many jobs as would be created by the construction of the proposed Boston Edison

Pilgrim II plant — a $1.5 billion project. Again, the American Institute of Architects has calculated that an all-out drive to make new and existing buildings energy-efficient could save 12.5 million barrels of oil each day by 1990 (nearly one-third of what America now uses) while employing between half and two million workers.

My Canadian correspondent emphasised that "the nuclear industry is well aware of the hazards involved in the peaceful use of nuclear energy"[34] — an oft-repeated statement that one hopes is correct. But it certainly does not wish the public to become equally aware of those hazards.

In Holland during March 1972, a senior technician at the Dodeward LWR, Theo Van Wass, felt bound to publicise the fact that numerous radiation leaks had been concealed. Inevitably he was sacked, whereupon several other technicians resigned in support of his action. Two-and-a-half years later an AEC safety officer, Carl Hocevar, resigned because of the extent to which the public was being deliberately deceived about the consequences of "minor" nuclear accidents. Then there was the celebrated occasion upon which the AEC tried to put pressure on Dr Tamplin to "amend" his radiological findings. Usually such pressure is not merely verbal and experts of less integrity sometimes respond to it.

On 3 April 1976 the *Economist* reported that the entire US geothermal research budget for the previous year was *less* than the bribes — $46 million — paid by the Exxon Corporation to Italian politicians who were hesitant about playing the nuclear game.

On 19 June 1980 the *Daily Telegraph* reported that Mr Wedgwood Benn, the Labour Government Energy Secretary, in evidence to the Commons Energy Select Committee, had "called for a Freedom of Information Act, to permit the maximum possible disclosure about nuclear power matters, so that people would know what pressures had been put on the Civil Service and Ministers to take certain decisions". Mr Benn said: "Without full disclosure, the many powerful interests which carry great weight with Whitehall will not become known and their effect in influencing decisions cannot be studied and assessed. For instance, the

Shah of Iran offered to invest in the British nuclear power industry if Britain dropped the gas-cooled reactor in favour of the American-designed water-cooled one."

During the Watergate investigation, Gulf, Ashland Oil and the Northrop Corporation, whose subsidiaries are among the main suppliers of nuclear technology, were found guilty of illegally "contributing" more than $10 million to politicians in America and abroad. Some of their beneficiaries were then influential members of the US Congress Joint Committee on Atomic Energy. Also, the top civil servants within ERDA and the NRC are, and always have been, "interested parties". In February 1976 it was disclosed that half of ERDA's senior men had previously worked with corporations involved in energy development and that more than half of the NRC's workforce had been recruited from universities, laboratories or corporations under its own supervision. The corporations included Rockwell International, Du Pont and General Dynamics — who need favourable decisions if they are to continue to profit from nuclear expansion — and America's "Big Five" reactor suppliers: Westinghouse, GE, B&W, General Atomic and Combustion Engineering.[35]

Industrial incest is nothing uncommon, but in the nuclear industry it can physically endanger the public. The Kemeny Report was unsparing: "When the NRC was split off from the old AEC, the purpose was to separate the regulators from those who were promoting the peaceful uses of atomic energy. We recognise that the NRC has an assignment that would be difficult under any circumstances. But, we have seen evidence that some of the old promotional philosophy still influences regulatory practices of the NRC. While some compromises between the needs of industry and the needs of safety are inevitable, the evidence suggests that the NRC has sometimes erred on the side of the industry's convenience rather than carrying out its primary mission of assuring safety." All this, and much more of the same, explains why anti-nukes now regard some "expert opinion" as unreliable.

The anti-nuke fear of plutonium is often ridiculed. Mr Howieson wrote: "The dangers of plutonium toxicity surely must be related to its availability . . . Dr Cohen has issued a challenge that he will eat as much plutonium as any anti-nuke will eat caffeine, secure in the knowledge that caffeine has the greater toxicity. The only route by which plutonium has the extreme toxicity mentioned — estimated thirty-year mortality — is via inhalation to the lungs. In my professional judgement, the industry and its regulators do recognise and guard against that possibility!"[36]

One feels some anxiety on behalf of the ill-informed Dr Cohen; let's hope he was only bluffing. As for my correspondent's "professional judgement", he is evidently unaware — being perhaps too absorbed in nuclear physics to read the newspapers — that the industry and its regulators, who are only human, *cannot* guard us against occasional disastrous releases of plutonium. When two tons of plutonium were accidentally burnt at a military reactor site at Rocky Flats, Colorado, escaped plutonium oxide reached parts of Denver, sixteen miles away. Radiation tests showed that thousands of acres of land, including a major water source, were contaminated. Four years later, in 1973, at hearings held by officials of the Colorado Department of Health, it was revealed that many local farm animals were being born with grotesque deformities. Whatever precautions are taken, the "availability" of plutonium must inevitably increase if we continue to create, store and transport this deadly element. There is ample scope for nuclear accidents but *no* scope for error. To whitewash catastrophes is relatively easy, since most nuclear accidents have long-term effects. But by the end of this century there may be many more witnesses to the invincible power of the god of the underworld.

Edward J. Gleason was one of the first witnesses. A New Jersey truck dock worker, he handled an unmarked, leaking container of chemicals contaminated with plutonium-239 on 8 January 1963. This defective box had spent five years lost near Pittsburgh; when handled in Jersey City it had been recovered and was on its way to the Brookhaven Laboratory on Long Island. Three years later Gleason developed a rare form of cancer. In 1968 his arm was amputated but the cancer continued to spread. Each year less than one person in a million develops such a cancer in the US, yet the transport companies argued that there was no conclusive proof that leaking plutonium had caused this particular cancer. Gleason died before the suit was decided.

A macabre thriller could be based on another infamous error,

The Case of the Missing Cask. This cask, containing 385 kilos of plutonium, disappeared en route from Ohio to St Louis and reappeared nine days later at Boston's Logan Airport. The amount of plutonium it then contained is not recorded. And these are only two of the many documented cases of American nuclear materials that went astray.

It is not known how much SNM has been "diverted" en route from one US facility to another but it seems reasonable to assume that thefts do occur at fairly regular intervals. In April 1969, at a meeting of the Institute of Nuclear Materials Management, Samuel Edlow, consultant on the transportation of nuclear materials, warned: "The transportation industry is so thoroughly infiltrated by the Cosa Nostra that any cargo which organised crime determines to obtain will be obtained. The environment of the transportation industry is one of incompetence, criminality and unreliability."[37]

Transport "incidents" are also common. On 20 December 1973 there was a traffic pile-up involving two cars, six trucks and a seventh truck carrying radioactive cobalt. One truck was carrying inflammable lacquer and one car exploded into flames; but the nuclear truck escaped being fired. On 12 January 1976 another road accident smashed eight 55-gallon drums of low-level waste and their contents were spilled on to a bridge over the Kentucky Highway. Much emphasis was put on this waste being *low-level*. Pro-nuke experts smiled in a carefree way on television while assuring the public that everything was just fine, because all that spillage was just low-level stuff, see, of no real importance . . . On 17 January 1979 twelve barrels of uranium oxide were dumped on to Interstate Highway 70, near St Louis, when the floor of a tractor-trailer gave way. For half-an-hour both lanes were closed while one was being cleaned; then traffic was allowed through while the second lane was still blocked. But many vehicles had quickly turned around and gone another way. By 1979 the public was becoming somewhat wary of industry reassurances.

The AEC's Hanford waste disposal facility, only seven miles from the Columbia river, is notorious for making plutonium "available". Over half-a-million gallons of high-level radioactive liquid have leaked from steel tanks buried a few feet below the surface. These wandering isotopes include plutonium, strontium-90 and caesium-137, all of which are equally deadly in their various ways. Also at Hanford masses of radioactive waste were regularly dumped into earthen pits because it was imagined that the contaminants would be absorbed by the soil and permanently held above the water-table in an insoluble form. An accumulation of plutonium, equivalent to at least ten Nagasaki-sized bombs, was thus allowed to collect in Trench-Z9. Eventually the AEC discovered that different elements were being absorbed at differ-ent rates, with resulting separate layers. One layer of concen-trated plutonium aroused fears that a nuclear chain reaction might be triggered by heavy rain, sending a mud volcano of radioactive wastes into the air — a repeat of Russia's Urals tragedy. At that point Congress provided $1.9 million to build an automated plutonium mine to cope with the crisis. But mean-while, presumably for lack of any alternative disposal system, the AEC was continuing to dump plutonium into a similar trench called Z18. This isotope is not amenable to quick changes of plan necessitated by human error.

Nuclear power and nuclear weapons are so closely linked that no government will come clean about costs. Not that these matter much, at the deepest level. As Walter Patterson has written: "If, in some fashion, the 'hidden' costs of nuclear technology become part of the visible accounts, energy planning will be much better able to evaluate the various options available, in numerical terms. But to many the ethical issues will remain unambiguously ethical, whose costs as such are simply irrelevant."[38]

In the '50s Lewis Strauss, then AEC Chairman, predicted that nuclear energy would produce electricity at a price too cheap to meter; but at no stage did that dream seem likely to be realised. Since 1964 America's consumer price index has risen 77 per cent while nuclear plant construction costs have increased by over 1000 per cent. Anti-nuke interventions during licensing hearings have forced the NRC to insist on improved, and therefore far more expensive, safety systems, and on modifications to plant design which can cost millions of dollars. Another grave handicap is that minor repairs sometimes require a shutdown. A broken pipe at the India Point-1 plant, twenty-four miles north of New York City, closed the plant for six months during which time 1,300 welders — almost every certified union welder in the New

York and Westchester areas — were enlisted to cope with one pipe. In America nuclear workers may receive thirty times as much radiation as the general public, yet many of those welders received a six-months dose in a few minutes. Possibly because of the risks involved, skilled labour has become very scarce. Welders competent to deal with nuclear systems can earn up to $40,000 a year, working overtime at double or triple pay. Expensive overtime is also necessary to cope with the shortage of electricians, boilermakers, pipefitters and metalworkers. Moreover, because of the complex plant design, whole teams of skilled men are often left idle while waiting for other teams to finish their particular segment.

The unreliability of LWRs perhaps explains why their manufacturers have to spend so much on bribes. Instead of producing electricity at the expected 80 per cent capacity factor, America's large nuclear plants averaged only 54 per cent in 1975. Many plants are ageing more rapidly than was foreseen and producing less as they wear out. In middle-age they tend to develop corrosion and to accumulate deposits of highly radioactive solids (crud) in the cooling system. The oldest commercial reactor in the US is Commonwealth Edison's Dresden plant in Morris, Illinois, which by the age of seventeen could not be repaired using normal methods. A replacement would have cost over $200 million so the owners planned a clean-up for $35 million, using untried methods hazardous to many workers. If "cruditis" cannot be cured, reactors "die" prematurely and the funeral costs are high.

Every plant should be decommissioned when it becomes too hot to handle and the average plant life will probably be about thirty years — with luck. The US General Accounting Office (GAO) has reported: "There are not firm estimates of decommissioning costs for large-scale nuclear power plants in operation or under construction. The seven reactors that have been decommissioned were small and differed substantially in design from modern reactors."[39] One of those seven was a mini-reactor built near Los Angeles in the '50s for $13 million and disassembled by remote control: which two-year task cost over $6 million. A burial ground for the plant's remains had not yet been chosen when decommissioning began. Another tiny plant cost half a million dollars to "mothball" in 1973; it was only 3 ME. The Elk

River Power Plant in Minnesota (22 ME) cost about $6 million to dismantle. So what is likely to be the cost of decommissioning such modern giants as the 1065 ME Brown's Ferry reactor? PG&E foresees that the costs of decommissioning each of its Diablo Canyon plants could be over $70 million. The Virginia Electric and Power Co. fears dismantling costs of $150 million for each of its North Anna plants. And the MITRE corporation has predicted that the costs of decommissioning could equal the initial construction cost of the nuclear section of any given plant. Yet the NRC has optimistically estimated that the mothballing of giants need cost only $3 to $5 million, plus annual surveillance costs — for an indefinite period — of $60,000 to $100,000.

"Mothballing" is the simplest way of decommissioning a reactor. It merely involves removing the coolant and fuel, then locking and guarding the plant. But this method is unlikely to work for reprocessing plants, which are heavily contaminated with plutonium and other transuranium elements. These installations will probably require "entombment" — filling or surrounding radioactive components with concrete — a process with the advantage of reducing the size of the permanent guard force. However, some experts argue that such highly radioactive structures could only be safely dealt with by "dismantling" — demolishing the entire plant, disposing of it as radioactive waste and restoring the site to its original state. (Someone wearing rose-tinted spectacles thought of that last item.) The snag about dismantling is that it simply shifts a highly contaminated mass from one place to another, possibly endangering the public in the process.

In a blood-curdling paper,[40] the New York Public Interest Research Group (NYPIRG) has pointed out that the nuclear industry's decommissioning calculations (such as they are) seem to overlook the possible presence in reactor "corpses" of the isotope nickel-59. This is formed as a by-product of the structural material in the pressure vessel. Atoms in the steel can absorb excess neutrons from the fission process and if a neutron is absorbed by an atom of nickel-58 or cobalt-59, the resulting isotope can be nickel-59 — which is radioactive with a half-life of 80,000 years. It would therefore take 1.5 million years to decay to comparatively safe radiation levels. Which means that decommissioned reactors, whichever "last rites" are used, could present no less of a problem than high-level waste.

This NYPIRG paper reinforced a slightly earlier report by the Energy Policy Project, a two-year study of national energy issues funded by the ubiquitous Ford Foundation and directed by S. David Freeman, former Director of the Energy Policy Staff of the Office of Science and Technology, a Federal government agency. The Project was blunt on decommissioning: "Present plans call for industry to turn over reprocessing plants to the states in which they are located, after they are taken out of service. But the states have no idea of what they will do with them. A basic difficulty is that reprocessing plants and nuclear reactors are not designed to be decommissioned. The radioactivity hazard is generally worse in reprocessing plants, but it can be a special problem in a reactor which has had an accident, as in the case of the Fermi breeder reactor in Detroit. A full assessment of the decommissioning problem should be carried out — promptly — before the new reprocessing plants coming on line are fully contaminated, and before reactors proliferate throughout the country. Institutional and economic questions are at least as important as technical ones. Who should be responsible for decommissioning? How should decommissioning be paid for? How will decommissioning effect the economics of the nuclear fuel cycle?"[41] (In October 1966 the Fermi FBR suffered a partial [four of its fuel assemblies] meltdown and the NRC has estimated its decommissioning costs at $7.1 million.)

By now nothing to do with the NRC should surprise us. Yet it does seem a little odd, even by NRC standards, that plant licensees are not required to provide decommissioning plans until a reactor has "died".

In Britain the UKAEC gave equally little thought to the morrow when initiating their Magnox programme; but now the CEGB and the SSEB annually put aside certain sums for de-commissioning— in 1978-9 the sum was £20 million.[42] Within the next twenty years, twenty-six Magnox reactors will require a decent burial and the AEC has devised experimental last rites for the small Windscale AGR, closed in March 1981. This decom-missioning will take place in three stages. First, all coolant and fuel will be removed; second, all structures outside the primary containment area and biological shield will be removed; third, the rest of the reactor will be removed — in theory within five years of shutdown. The AEA has mentioned £20 million as the cost of stages two and three.[43]

Everywhere the nuclear industry refuses to accept full liability for a major nuclear accident. In America the Price-Anderson Indemnity Act was engineered in 1957, to aid the development of civil nuclear power.It guarantees $500 million from Federal funds, in addition to the $60 million made available from private insurers. Here again the tax-payer is propping up the industry. And in Britain, too, the government has rallied round. According to the Nuclear Installations (Licensing and Insurance) Act 1959, the limitation on operator liability is established at £5 million from private insurers, plus £43 million (raised to £45 million by a later Act) from the government. As Walter Patterson has pointed out: "Since other industries, including other energy-supply industries, must provide their own third-party coverage out of working funds, there is little doubt that the nuclear industry gains a distinct competitive advantage from the present provisions."[44]

Ralph Nader and John Abbott have commented: "The NRC and the industry continue to claim that the chances of a reactor accident are very slight: one in a million per year or less. So the question which must be put to the industry is: If the probability of a reactor accident really is so low, why will the industry not accept full liability for such an 'incredible' accident? If the industry statements were, in fact, true, then the industry would be risking very little by accepting liability for an accident that, practically speaking, will not occur. But the fact that the nuclear industry continues to demand the protection of Price-Anderson indicates that it believes a catastrophic accident to be a distinct possibility."[45]

On 17 September 1979 a letter was quoted in the *New York Times* which had been written three years previously, to Mr Carter's campaign policy committee, by John F. O'Leary, a former Deputy Secretary of Energy and AEC staff official. Mr O'Leary argued: "The frequency of serious and potentially catastrophic nuclear accidents supports the conclusion that sooner or later a major disaster will occur at a nuclear generating facility. The NRC . . . has been unwilling to face up to the policy consequences of assigning a high probability to a serious nuclear accident." The publication of that letter was prompted by the Indian Point controversy, then at its height.

Nucleus reported: "On September 18, the New York State Department of Energy announced that it was asking the NRC to reopen hearings on the need for earthquake monitoring equipment at Indian Point. Currently, no such monitoring equipment is in place even though the plants are sited less than one mile from the Ramapo fault which, though quiescent, has been judged to possess the potential for movement during the operating lifetime of the Indian Point power station."[46]

This station is only thirty miles north of New York City, in Buchan. The UCS believes that there is "a clear and present danger" to the nineteen million people — almost 10 per cent of the population of the US — living within a sixty-mile radius of the site. Therefore in September 1979 these distinguished scientists — feeling even more concerned than usual — announced the filing of a legal petition with the NRC requesting the closure of Indian Point II and III, to give the Federal government time to consider the consequences of a major accident at either plant. The petition also requested immediate repairs to gross safety deficiencies at both plants. And it asked that Indian Point I, lying idle since 1974, be deprived of its infamous *seventeen-year-old* "provisional" licence and decontaminated and decommissioned without delay. Indian Point I had been idle for so long because the owners, Consolidated Edison Co. of New York, were reluctant to install basic safety equipment ordered by the NRC. (As Indian Point I went critical in *August 1962*, one wonders why the NRC ordered *basic safety equipment* only in *1974 . . .*) Con Ed were also reluctant to incur the unknown but huge costs of decommissioning, though as we have seen a defunct reactor is *not* a harmless bit of rubbish.

Indian Point II and III were designed as twins but are not identical. During the safety review of III, the NRC ordered major design defects to be corrected — but, inexplicably, II was allowed to operate *with* those defects. Moreover, the NRC itself decided in 1978 that the fire protection in both plants was inadequate. Yet the reactors were allowed to continue operating, despite the fact that a fire could simultaneously put all primary and back-up safety systems out of order, thus causing a meltdown. No wonder the UCS took action.

Nucleus explained: "According to the 1975 *Reactor Safety Study* — repudiated for its conservative bias by the NRC in early 1979 — a meltdown at a typical nuclear plant could kill as many as 48,000 people from acute radiation exposure and radiation-induced cancer, cause another 285,000 nonfatal nuclear radiation injuries, give rise to 5,100 genetic defects in the first generation born after the accident, result in $14 billion in property damage, *and* contaminate 3,200 square miles of land with such high levels of radioactivity that it would remain uninhabitable for a century. Indian Point is not a typical power plant site, however. These reactors are in the midst of an uncommonly large number of people who could not be rapidly evacuated in the event of a meltdown. (Indeed, the *Reactor Safety Study* itself stated that for New York City and other metropolitan areas, 'there is no presumption that the population could be moved in less than one week'.) The unsuitability of the Indian Point site is widely acknowledged. Charles Luce, chairman of Con Ed, the electric utility responsible for siting the plants only thirty miles from New York City, has acknowledged this, admitting, 'I doubt that we would propose to build Indian Point at that particular location if we were building it today.' "[47]

Considering the foregoing, it is hardly surprising that Herbert Denenberg, a former Pennsylvania insurance commissioner, has estimated the normal annual premium on a commercial reactor at not less than $23.5 million — about what it costs to operate and maintain a plant for one year.[48] The nuclear industry could never have got started without the Price-Anderson Act and nukedom's shaky economic foundations are regularly exposed in the American press.

A classic case concerns America's first privately owned nuclear fuel reprocessing plant at West Valley, New York. A well-known nuke pioneer, Nelson Rockefeller — then Governor of New York — solemnly broke ground for this plant in 1965, remarking that it would "make a major contribution towards transforming the economy . . . of the entire state".[49] Instead, it may eventually make a major contribution towards transforming the health of a considerable percentage of the state's population. Within nine years, thirty-eight people on the site had either ingested or inhaled "excessive concentrations of radioactive materials"[50] in at least fifteen separate "incidents". And these unfortunates represented only the tip of an exceptionally nasty iceberg. "The plant's practice of using young transient workers from the area's large pool of unemployed people for particularly 'hot' jobs — where they would receive their annual or quarterly dose of radiation in minutes or hours and be sent home with a day's pay — undoubtedly condemned many innocent people to future cancers."[51]

This $32.5 million disaster was built and initially operated by Nuclear Fuel Services (NFS), a subsidiary of W. R. Grace-Davison Chemical Co., which in 1969, after operating for three years at a loss, sold the plant to Getty Oil. (The recovered plutonium was used for nuclear weapons.) Reprocessing, still at a loss, continued until 1972, when the plant was closed. However, West Valley functioned (or malfunctioned) as a high- and low-level waste store until the spring of 1975, when Getty Oil announced a complete closure. Meanwhile, radioactive materials from the plant had been released into Cattaraugus Creek, and buried radioactive wastes had leached into that same creek, which runs into Lake Erie, the source of Buffalo's municipal drinking water. When Getty Oil explained that the closure had been necessitated by changed regulatory requirements, which demanded an additional investment of $600 million (almost twenty times the plant's original cost), it seemed to many that those changed requirements were somewhat belated ... At this stage, the misdeeds of the Windscale management begin to look like pecadilloes.

The original Rockefeller-approved contracts between New York State and NFS left final responsibility with the state for the containment and disposal of wastes, and for decontamination and decommissioning of the plant. NFS was required to establish a reserve fund for the purpose but in 1979 there was less than $4 million in that account; and ERDA reckoned that merely to solidify and store the 600,000 gallons of high-level wastes, left in temporary underground tanks, could cost $500 million. (This is to ignore the two million cubic feet of buried low-level garbage.) A 1976 Federal government study warned that on average the West Valley site could have a severe earthquake every 750 years — but what has posterity done for Getty Oil? According to ERDA, decommissioning might cost $1 billion. Of course that has to be a guess, as no such plant has yet been decommissioned anywhere in the world. (What will it cost to decommission Windscale? Has the AEA attempted an estimate?)

While our old friend ABCDG argues about West Valley's lethal deposit, the tax-payers of New York State are paying between $2 and $3 million a year to maintain the alarming *status quo*. The Federal government has been asked by Governor Carey to help foot the bill left behind by NFS when they were conveniently declared bankrupt. And the Governor has declared that he will veto all nuclear construction in the state until the waste problem has been solved.[52]

Perhaps anti-nukes should conserve their energy; the Mad Hatter economics of the nuke industry seem likely to bring about its demise, sans anti-nuke campaigns. As Anna Gyorgy has put it: "Much of the battle over nuclear power can be interpreted as a battle over who will bear the *real* costs of nuclear power. When reactor manufacturers were forced by an AEC decision to build reactors which released less radiation, the price of reactors — and thus of nuclear-generated electricity — went up. Instead of local residents bearing a non-monetary cost, the increased risk of cancer, utilities were forced to bear the cost of a safer reactor. When utilities are forced to take environmental impacts into account, the improvements save lives but cut profits."[53]

For years the construction of expensive nuclear plants enabled electricity utilities to invest billions of dollars and guarantee ever-increasing profits to shareholders — provided by helpless customers with no alternative but to pay up. Since the mid-70s, however, the universal handicaps of inflation and rising wages have been aggravated for nuke-owners by the uranium price-rise and the inability of many components manufacturers to deliver orders on time. Cost overruns have seriously threatened profits. In 1975 the following plants were completed and I give their cost overruns in brackets: Calvert Cliffs (58 per cent); Brunswick II (91 per cent); Hatch (149 per cent); Cook I (80 per cent); Millstone Point II (125 per cent); Trojan (98 per cent); Brown's Ferry I and II (168 per cent).[54]

Boston Edison officials have always been eloquent about the cheapness of nuclear power, yet in 1975 they raised their rates by $29.5 million and a year later by another $11 million. Early in

1979, Connecticut's Northeast Utilities requested an electricity rates increase of $131.2 million to enable them to pay their share (65 per cent) of the Millstone II reactor. They also wanted $7.7 million to pay for costs incurred while Millstone was closed for repairs in 1978. The argument is that since customers have benefited so much from "cheap" nuclear electricity they should be happy to pay for these little incidental expenses. Northern Utilities is also attempting to force customers to pay increased rates whenever their plants operate below 67 per cent capacity. Worst of all, this utility wishes to charge users of small amounts of electricity at 33 per cent increase, while charging others "only" 14 per cent. This provocation to waste still more energy is designed to make new generating capacity seem essential. To the nuclear industry, "energy conservation" is a dirty phrase. Occasionally spokesmen make hypocritical remarks about saving energy but their propaganda is geared to persuading people to equate such saving with a general lowering of the standard of living.

Looking at the controversy from another angle, to what extent can nuclear power truly ease the oil-shortage? Some surprising figures were given in a *Foreign Affairs* article during the summer of 1980:

Replacing oil is undeniably urgent. But nuclear power cannot provide timely and significant substitution for oil. Only about a tenth of the world's oil is used for making electricity, which is the only form of energy that nuclear power can yield on a significant scale in the foreseeable future. The other nine-tenths of the oil runs vehicles, makes direct heat in buildings and industry, and provides petrochemical feedstocks. If, in 1975, *every* oil-fired power station in the industrialised countries represented in the OECD had been replaced *overnight* by nuclear reactors, OECD oil consumption would have fallen by only 12 per cent ... In practice, US nuclear expansion has served mainly to displace coal, not oil, by running coal-fired plants less of the time: the utilisation of their full theoretical capacity dropped from 62 per cent to 55 per cent during 1973–8. In overall quantitative terms the whole 1978 US nuclear output could have been replaced simply by raising the output of partly idle coal plants most of the way to the level of which they are practically capable ... Between the first quarters of 1979 and 1980, total US oil-fired generation fell 32 per cent while nuclear output simultaneously fell 25 per cent — hardly a substitution ... The example of Japan, widely considered the prime case of need for nuclear power, illustrates reactors' relatively small eventual contribution to total energy supply. Quadrupling Japan's nuclear capacity by 1990 would reduce officially projected oil import dependence by only about 10 per cent . . . The official projections reflect an inability to face the fact that nuclear power cannot physically play a dominant role in any country's energy supply. Solving the oil problem will clearly require, not a nuclear panacea, but a wide array of complementary measures, most importantly major improvements in energy efficiency.[55]

My Canadian adversary unwittingly underlined the futility of nuclear power. According to Mr Howieson: "The sheer scale of the present energy supply systems are (*sic*) not understood by most people. The world's energy supply has been increased by the existing nuclear plants by approximately 2 per cent for an investment already of $200 billion."[56] As the West's percentage of wasted energy is closer to 22 per cent than 2 per cent, those $200 billion could have been invested in alternative energies r & d without any lowering of our living-standards. On this subject Mr Howieson gets into rather a muddle: "The widespread adoption of biomass and solar energy is being encouraged but, unfortunately, is only likely to be widely adopted when large-scale manufacturing brings the price down. It might also be pointed out that the countries with the most need for energy — the northernmost of the developed countries — will be immediately at a significant economic disadvantage, for a given amount of energy based on the sun will cost such countries much more to extract."[57] This last point may well be true. Yet one is not ravaged by anxiety on behalf of "the northernmost of the developed countries", who can well afford to be at a disadvantage.

Having expressed his distaste for nuclear weapons, Mr Howieson looked on the bright side: "It would not be the first time over-abundant life has been wiped out on earth. Such a fate overtook the world in the time of the dinosaurs but life survived and diversified."[58] Has no one ever explained to nuclear scientists the sort of diversification likely after a nuclear war? Or is "nuclear numbing" for them an essential protective device? Dr John Gofman has noted: "The technical servant of the privileged élite does have a slight problem. Commonly of peon origins and commonly of fairly high intelligence, at least with respect to handling mathematics, physics or chemistry, there is always the danger that he will suffer psychological difficulties in serving anti-human goals while carrying through the plans and programmes of his privileged masters, be they commissars or capitalists."[59] The minority who cannot overcome those "psychological difficulties" resign from the industry: the rest allow themselves to be convinced that they are the saviours of twenty-first-century mankind.

Mr Howieson described himself thus: "I am one of those individuals who knows that the world's supply of 'conventional' energy resources is going to run out, and for the past twenty years have dedicated myself to providing an alternative — namely, nuclear power. It seemed to me to be morally right to turn swords into ploughshares and the nuclear industry has done just that. Most of us have been attracted to the business by that same moral objective . . ."[60]

For scientists of Mr Howieson's generation, it must be hard to accept the fast-accumulating evidence that fissile ploughshares retain many of the lethal characteristics of the original swords. But the younger generation seem to have got the message and are much less interested in nukes than in other applied science courses. On the telephone from Sweden, in July 1980, a professor of nuclear physics admitted to me that only two students are taking his 1980–81 course. And his colleagues in West Germany, Switzerland, Italy and the US have similar complaints. By now, in the minds of many of the American and Continental young, nuclear power is inextricably associated with duplicity and bullying. And increasing numbers of Britain's young associate it with governmental misspending, military arrogance, sneaky health hazards and blinkered planning.

Opponents in the nuclear debate tend to lose sight of each other's humanity. Many pro-nukes regard anti-nukes as dangerous dissidents: many anti-nukes regard nuclear scientists as faceless monsters doing diabolical deeds in carefully guarded buildings to which the general public may never penetrate — either physically or intellectually. So it was salutary to read Mr Howieson's simple, sincere statement of the pro-nuke case, seen from the point of view of an ordinary, kindly, hard-working scientist. It is not the fault of such individuals that "Atoms for Peace" has been proved, in David Rosenbaum's words: "One of the stupidest ideas of our time."[61]

5

How Irradiated Can You Get?

It does not make any sense to replace the present energy crisis with a long-term radioactivity crisis.

RALPH NADER[1]

At Salt Lake City my Greyhound bus from San Francisco to Chicago stopped soon after midnight for maintenance. I walked briskly around Temple Square, well wrapped-up against a sub-zero wind; the air tasted of snow and above me towered the sharp, symmetrical spires of the Mormon Temple, silhouetted against a sky of sparkling stars. On my way back to the bus station I was joined by the only other perambulating passenger, a burly, middle-aged man also wearing an anti-nuke badge. When I remarked that anti-nukes seem to be quite rare in our generation he replied, "Not around here!" And during the next stage of the journey he explained why, as he put it, the people of Utah "have a high level of nuke-consciousness".

In Nevada, between 1951 and 1958, the American Army conducted at least ninety-seven above-ground atomic tests — which were postponed if the wind was blowing towards Las Vegas or Los Angeles, but not if it were blowing towards the thinly populated areas of Utah and Nevada. In 1953, 4,300 sheep which had been on pasture downwind from the test-site — some as far as 120 miles away — were killed. They had suffered radioactive iodine doses up to 1,000 times greater than the maximum allowed to the human thyroid and also suffered beta-burn lesions on the ears, neck and face. Apart from the original atom-bomb victims, these animals and the human victims of the Pacific H-bomb tests are the only recorded cases of beta-burns. Ewes died either during lambing or shortly after and most of the lambs were stillborn. Yet the AEC asserted that those sheep had died of natural causes.

According to Dr Stephen Brower, a Brigham Young University professor, the Federal government attempted to hush up the ranchers by offering to finance a desert ranch nutrition research programme. At the time, Dr Brower was a county educational officer for the affected regions.[2]

In 1974, when the Utah topsoil was found to have a plutonium level 3.8 times higher than anywhere else in the US, the locals first realised that for over twenty years they had been exposed to an abnormally high level of radiation. Shorter-lived radioactive elements were also detected and nobody was reassured by the AEC's soothing remarks about those elements becoming less radioactive every year.[3]

When the 1,350 square miles Nevada test-site was chosen, official documents described the area as "relatively uninhabited". The population of some 28,000 strict Mormon farmers led outdoor lives, were uncontaminated by nicotine or alcohol and scarcely knew what cancer was. But thirty years after the first test most families can list several cancer victims and Stewart Udall, a former Secretary of the Interior, is sueing the Federal government on their behalf. Granted, many Americans are ready to claim damages at the drop of a hat. But in Nevada something more than a hat was dropped.

To quote Mr Udall: "The Federal Government has created a new class of people — 'relative uninhabitants' . . . In the early days, the AEC made a quick judgement that fallout did not cause health damage. But the only thing they used as a measuring stick was "external exposure" — the dust which fell on people. They didn't calculate what would happen when it got into the food chain or when it was inhaled. They simply assumed that if you didn't get radiation burns, your body may have suffered some insult but it would recover. Later, when evidence began to show up to the contrary, they ignored the evidence."[4]

In 1953, in upstate New York, heavy rain brought down abnormal fallout concentrations which came to be associated with an abrupt rise in childhood leukemia deaths in the cities of Troy and Albany. In 1962 the Ministry of Health, Alberta, Canada, detected a sudden increase in the local rate of congenital malformations and linked this to contaminated rainfall over certain areas: e.g., a 78 per cent increase followed the 1958 Russian bomb-tests.[5]

On 1 March 1954 the Bikini Atoll test of the first H-bomb — which proved to be twice as powerful as expected — irradiated the crew of an American destroyer and the inhabitants of the nearby Pacific islands of Uterik, Tongelap and Rongerik. The islanders later developed anorexia, various blood diseases and cancers, a grossly abnormal incidence of miscarriages and a 90 per cent incidence of thyroid tumours among children who were under the age of puberty in 1954. (A US government survey found the islands still heavily contaminated in 1978.)

The unhappy fate of a few hundred American sailors and Pacific islanders would probably have passed unnoticed by the world. What made things awkward for the bomb-testers was the coincidental proximity of a Japanese fishing-boat most inappropriately named *The Lucky Dragon*. Its crew of twenty-three were fishing for tuna some eighty-five miles east of Bikini when the bomb went off. A fortnight later they and their craft reached port sizzling with radioactivity and at once the world was alerted — not only to the coming miseries of sailors, islanders and fishermen, but to the *global* threat of fallout. Public opinion soon forced the declassification of secret information and the study that followed concluded that atmospheric bomb tests were causing, and would continue to cause, major and widespread genetic mutations. Yet these tests went on for another *nine* years, during which the International Commission on Radiological Protection (ICRP) made no protests to the governments concerned and did or said nothing to arouse public awareness of the unique hazards involved. The ICRP was established in 1928 and is the authority quoted by the nuclear industry when it wishes to quell public unease. An unofficial body of twelve men (in 1976 four were British), it is elected every four years by the International Congress of Radiology (a professional gathering) on the basis of individual scientific reputations. It recommends basic standards for radiological protection which are accepted all over the world. Flowers noted (para. 203): "The ICRP recommendations are expressions of opinion based on deductions from scientific fact. Clearly the ICRP is only as good as its members, and it is vitally important that these should continue to be appointed independently of the approval of their national governments and purely on the basis of their professional standing . . . Given this, we can see no better way of deriving basic standards than by accepting the ICRP recommendations."

Ralph Nader and John Abbotts have explained: "Historically, radiation standards have been continually made more stringent as more information has become available. This is illustrated by the proposals of the ICRP and the US National Council on Radiation Protection (NCRP). In 1925, recommended workers' limits allowed up to 100 rem per year, a standard which might have prevented immediate death, but which did not recognise radiation's ability to cause cancer or genetic effects. In 1934, the NCRP recommended 36 rem per year, which in 1947 they lowered to 15 rem per year for workers. In 1958, the recommended average yearly occupational dose was reduced to 5 rem per year, which is presently in effect. However, both the historical record and the continuing controversy on radiation standards suggest that further reductions may be in order."[6]

In Britain, "government surveys of those most exposed to radiation have consistently been incomplete, and therefore any conclusions drawn concerning the numbers of cancers caused, if any, must be viewed with great suspicion. It is of course convenient to the industry that failure to collect such evidence enables it to claim that there is no reason to suppose that harm results from their activities. A recent report on nuclear establishments in Britain by the UK Health and Safety Executive typifies these failings: thus it states that not one of the 7,500 workforce at 11 nuclear power stations received more than the internationally recommended upper limit of 5 rems per year in 1977 or 1978. Yet the survey failed to take into account some 1,500 temporary workers brought in by the generating boards to work in high radiation areas. Many of these temporaries carried out delagging operations on radioactively contaminated pipes, and received their quarterly maximum dose of 1.5 rems in less than 2 weeks. They were then taken off the job and replaced. Most important from the survey point of view, they are not being followed up for possible evidence of radiation-induced disease. The misleading practice of 'burning out' casual workers and replacing them, without incorporating them into statistics, is one increasingly followed throughout the nuclear industry as its need for 'cannon fodder' grows."[7]

In France this use of casual labour is even more blatant. Small employment agencies have been established in the towns nearest nuclear installations, to provide men who work by the hour or the day in contaminated areas. Their dosages are the responsibility not of the plant but of the agency, which makes no enquiries about whether or not these *intérimaires* have been exposed to radiation at other installations. The notoriously defective Cap La Hague reprocessing plant is particularly dependent on *intérimaires* "who may be asked to do anything. In the course of a few days they may receive as much radiation as a regular employee receives in a whole year. Sometimes they actually receive much more, for the agencies that have hired them often simply 'forget' to send in the check films required by the health authorities on which the daily dose received can be read ... Most of those who are engaged in this way are unemployed and are merely told they will be well paid, not how dangerous and responsible their work at La Hague is going to be. And they will have a tendency, not understanding the extent of the dangers they face, to take the safety precautions lightly."[8]

The British National Radiological Protection Board (NRPB) was requested by the Royal Commission to review Dr Gofman's suggestion that the risk of cancer from inhaled plutonium particles might be much greater than previously supposed. But (para. 70): "Our call for detailed review of material that had not been published in a journal of scientific standing ... provoked a comment that such work was time-consuming and could be counter-productive in diverting attention from the real issues of plutonium dosimetry. We note here, however, that because of the emotive nature of the issues raised by plutonium, which concern many people who are unable themselves to assess the scientific arguments, much attention is focused on work that suggests that current assessments of the hazards are seriously wrong. It appears to us that work arousing serious public concern should indeed be examined and, if appropriate, publicly refuted on the basis of more serious arguments than the mere denigration of the authors."

Much later on, Flowers (para. 204) refers again to the Gofman controversy and observes that: "Such dispute is likely to be heightened if, as may happen, a change in a standard would require changes in operating practices that would have consider-able economic penalties. There is then the need to balance economic factors against uncertainties in scientific evidence." The implications of that last sentence are disquieting.

Since 1959, the permissible dose for the British public has been 0.5 rem/year. In the US it was 0.17 rem/year from 1958 to 1974, when it was reduced to 0.005 rem/year. Many American radio-logists are now debating when and by how much to reduce the permissible dose for nuclear workers. Yet in 1977 the ICRP recommended continuing the previous annual dose limit. In a report known as ICRP-26, they allocated different risk rates to different organs, if irradiated *on their own*. These have been misinterpreted as allowing higher levels of critical exposure — in fact most of the internal exposure levels have had to be reduced.

However, even a layman can see that the concept of different risk rates for different organs is absurd — a bit of abstract non-sense dreamed up by men who experiment on animals in labor-atories. How does one irradiate a liver or kidney or thyroid gland *on its own*, while it is still in daily use? What happens to the rest of the owner's body while this irradiation is taking place? As is often the way with high-powered specialists, some ICRP researchers concentrate on the use of abstruse equations and computer models to *estimate* the effects of radiation and seem indifferent to the results when their conclusions are tested in the crucible of real life.

ICRP-26 stated: "Medical surveillance has no part to play in confirming the effectiveness of a radiation protection pro-gramme." This announcement has also, and not surprisingly, been misinterpreted by non-scientific anti-nukes. Clearly medical surveillance of workers who are regularly exposed to radiation is extremely important. *But,* as one distinguished radiobiologist explained to me, by the time a surveillance team has detected radiation within a body it is too late to help the victim. Therefore surveillance can only confirm the *in*effectiveness of a protection programme. At present the permissible dose for workers and public includes *internal* radiation, which is increasingly being detected in significant amounts among workers.

No wonder the most frequently played anti-nuke card is radi-ation. We cannot see, hear, smell, taste or feel it; yet we know that it can kill us — in five, ten, maybe thirty years. One pro-nuke friend of mine has suggested that for the rational, educated, irreligious citizen of the late twentieth century, this new, mysterious peril replaces those spells, ghosts and wrathful gods that so unnerved our superstitious forefathers. Many anti-nukes, he believes, feel a deep need for the intangible threats of the past, now banished by scientific knowledge. Rejecting the monochrome certainties of our time, they seek elusive, unpredictable dangers and work themselves into a state of absurd anxiety about a hazard that is perfectly amenable to human measurement and control. So they have created the imaginary spectre of Radiation stalking the earth, though they remain unalarmed by far commoner and more immediate dangers — such as traffic accidents, pesticide pollution and chemical explosions.

There is some truth in all this: certain anti-nukes sound so paranoid about radiation that one cannot possibly take them seriously. And inevitably lay-people are scared and bewildered by the experts' incessant bickering about the precise effects of radiation on the human body. Yet we may expect nothing else for decades. Most radiologists are agreed that laboratory tests on animals have a limited value, while the tests on humans now being inadvertently conducted, among people who live near or work in nuclear establishments, will not quickly yield results. In 1968 the US set up a "Transuranium Registry" to check on the subsequent medical histories of workers exposed to plutonium and other actinides. But in Britian similar record-keeping is not permitted by the nuclear industry.

Specialist dissension allows distortions to proliferate. At an Energy Symposium held in Dublin under the auspices of the Irish Transport and General Workers' Union, in May 1978, John F. Carroll, then Vice-President of the Union, warned his listeners that if Ireland built a nuclear plant, which would have to be decommissioned after some thirty years, "the only jobs involved ultimately are for security guards patrolling the fenced-in decommissioned reactor and site. And these jobs would require the occupants to be dressed up in protective clothing so as to make them look like something out of a space-fiction magazine. I'm not sure that there would be many takers for those jobs, bearing in mind the hazards of radiation at the site."[9] This of course is nonsense and pro-nukes frequently seize on such exaggerations to prove how ill-informed we anti-nukes are. Although nuke "corpses" will have to be guarded for many centuries, against saboteurs and terrorists, protective clothing should not be necessary if the reactors have been responsibly decommissioned. (But is this a king-sized "IF", given the cost of decommissioning and the performance of the industry to date?)

Even those involved in the nuclear industry can suffer from serious misapprehensions. In London I talked with Sir Frederick Warner, a distinguished engineer who sat on the Flowers Commission. I asked him: "If the AEA can already safely store krypton-85, why not do so?" And he replied: "K-85 storage is fiendishly difficult and expensive. It can be done by ion implantation but is it worth the resources considering the reduction in skin cancer will be many times made up for by sunbathing?"

However, at the Hearing on Nuclear Energy organised by the World Council of Churches at Sigtuna, Sweden, in June 1975, it was emphasised in the summary (A Report to the Churches) that: "Considering all sources of radiation from a nuclear reactor system, including reprocessing, it is worth noting that releasing tritium and krypton-85 to the atmosphere may become the main source of radiation dose to the population from the nuclear power industry. Techniques exist at present for collecting krypton-85. Such techniques must be applied in future reprocessing plants. The technique for collecting tritium is more complicated, and it must be expected that fairly large amounts will be released into the environment, most probably in the form of water. The difficulty of retaining it may become a limiting factor for the nuclear industry."[10]

Sir Frederick is no unscrupulous tycoon who puts profits before people; he genuinely believes the krypton-85 hazard to be trivial. Considering the importance of matters nuclear to the present and future well-being of the whole of mankind, public confusion and ignorance on the subject is quite astounding: and for this the evasive and devious nuclear industry is greatly to blame. Expert ignorance about the long-term effects of low-level radiation is inevitable and nobody's fault. But the industry and its supporters must be criticised for attempting to conceal (and in some cases perpetuate) that ignorance.

At Cap La Hague — Western Europe's largest reprocessing plant, on the northwestern tip of the Cherbourg Peninsula — the graphite sheath of nuclear fuel waste caught fire in its silo on 6 January 1981. Fine dust then blocked a filter, forcing contaminated smoke through the opening used for charging the silo; but, according to the plant management, this smoke contained neither strontium nor caesium. However, a painter and nineteen firemen were contaminated; the painter and three firemen were sent to be checked at France's main centre for protection against radiation, at Le Vesinet, near Paris.

On 9 January the management asserted that none of the twenty men had been exposed to the slightest health hazard, which, as a *Times Service* report noted, "on its face value, seems a little contradictory". The staff trade union then accused the management of playing down the incident, refusing to give accurate information about its significance and neglecting to take the precaution of complete staff evacuation and subsequent radiation checking. The painter, they claimed, had received the maximum permissible radiation dose for one year. Yet the management repeatedly emphasised that the twenty men had "been contaminated in a very negligible fashion and without any consequences".[11] This is a classic example of industry bluff. It is scientifically impossible, within days of a man's being contaminated (however slightly) to determine the extent of the damage done to his body — damage which may not become apparent for twenty years or more.

In February 1981 the anti-nuke cause was greatly strengthened by the first official finding that long-term exposure to low-level microwave radiation can prove fatal. The case concerned Samuel Yannon, a New York Telephone Company supervisor who worked with television relay equipment for eight years. In 1970 he had to retire and he died four years later, of "abnormal, premature ageing", according to Dr Milton Zaret, a radiation specialist and professor of New York University.[12]

Nuclear accidents, however slight, are hazardous. Each minute release of radioactivity, by adding to the planet's natural level of background radiation, could eventually do harm. And no continent can escape the effects of a major nuclear disaster anywhere in the world; we all share the same biosphere, the same wind-currents, the same rain-clouds and oceans. (This radiation

crisis may do some good if it makes us truly aware of the brother-
hood of man, not just as a nice idea but as an actual, physical
reality.) Even now we are being affected by fallout and by the
many minor, and not so minor, accidental releases of the past
thirty years; releases which when they occurred were either
concealed or made to seem trivial. In 1975 the US National
Centre for Atmospheric Research revealed that more than five
metric tons of plutonium had by then been thinly dispersed over
the earth, as a result of bomb-testing, effluents from nuclear
reprocessing plants, explosions, leakages, fire and spills. Nor
can this contamination be checked, while the nuclear industry
continues to operate.

Contrary to popular opinion, the anti-nuke movement is not
led by long-haired vegetarian pot-smokers, of vaguely Marxist
inclinations, who wouldn't know a gamma from a beta ray. Many
reputable scientists are among its most fervent supporters —
including, in Britain, Professor Sir Martin Ryle, FRS, the present
Astronomer Royal, and Sir Kelvin Spencer, CBE, who was Chief
Scientist at the Ministry of Fuel and Power during the early years
of the civil nuclear power programme.

America's anti-nukes include Drs John Gofman, Thomas
Mancuso, Arthur Tamplin, Ernest Sternglass, Tom Cochrane
and Rosalie Bertell. These eminent scientists all disagree with the
US National Committee on Radiological Protection (NCRP).
The statistical data involved are open to a variety of interpret-
ations and the fallout from these clashes seems likely to irradiate
the medical world for many years to come. Of necessity (I suppose) this civil war is being fought almost entirely with statistics, a
weapon distrusted by every sensible person. And, since no non-
expert can assess the reputations of scientists, we must instead
assess their *motives*. Had all these anti-nukes grown up hostile to
nuclear power, they might be vulnerable to accusations of having
cooked their books. But in every case it was the other way round:
they started out happily associated with the nuclear industry,
then were turned against it by what statistics seemed to reveal.
Most of them have sacrificed secure careers to their principles.

For years past, one of the main radiation controversies has

raged around the work of Dr Alice Stewart, of the Department of Social Medicine, Oxford University. In 1958 she published a large-scale epidemiological study which suggested that the damage done by small doses of radiation had hitherto been seriously underestimated. She claimed that exposing children before birth to only a few diagnostic X-rays almost doubled their chances of developing leukemia and other cancers during their first ten years.[13] Twelve years later, Dr Stewart completed a study of some ten million children, born in England and Wales between 1947 and 1965, which revealed that for foetuses irradiated during the first three months of pregnancy the risk was much greater than for those irradiated just before birth.[14] Britain's once-popular mass-radiography programmes have since ceased.

Using Dr Stewart's figures, Dr Ernest Sternglass has argued:

One sees that the risk of childhood leukemia and cancer, from normal operations of nuclear plants, might be increased by 20 per cent or more for children born near such facilities, drinking the water and local milk, without the legally permitted doses being exceeded. However, recent studies have shown that the risk of cancer and leukemia is actually not the dominant one for exposure during intra-uterine development, since only about one in 1,000 children develop leukemia or cancer before age ten. A large-scale prospective study of mothers who received abdominal X-rays, in the course of necessary diagnostic exposures, carried out at John Hopkins University by Diamond, Schmerler and Lilienfeld, showed that an even greater increase in deaths per 1,000 births took place for diseases of the respiratory and digestive systems. Altogether, for all causes of deaths combined, this very careful study sponsored by the US Department of Health and Welfare showed that for the exposed group of white children, the mortality rate in the first ten years of life was 18.3 per 1,000 births, as compared with 9.8 for those who had not been exposed to radiation in utero. And once again, those who had been exposed at an earlier stage of development showed a much greater risk than those X-rayed just before birth. The indication of increased risk of infectious diseases, or reduction in the effectiveness of immune-system defences, following intra-uterine exposure to small amounts of radiation, is perhaps the most serious potential health hazard of environmental radiation ... Not only are viruses and bacteria able to multiply more readily when the body's immune defences are impaired by small amounts of radioactivity in the food and water — the same has been found to be the case for cancer cells, which are normally destroyed by the action of certain white blood cells or phagocytes. Thus, low-dose radiation, by reducing the ability of the body to destroy individual cancer cells before they can multiply out of control, can also lead to much greater risk of cancer than had been expected on the basis of the high-dose, short duration exposure experienced by the A-bomb survivors on which present radiation protection standards are based.[15]

Anti-nuke scientists occasionally disagree among themselves, which seems healthy. Dr Arthur Tamplin found Dr Sternglass's estimate of US infant deaths resulting from the Nevada bomb-tests much too high: he himself concluded that those tests had killed some 4,000 infants — rather than the 40,000 suggested by Dr Sternglass. However, Dr David R. Inglis, a distinguished physicist who since Manhattan Project days has worked within the nuclear industry, analysed Sternglass's controversial study (assisted by Dr A. R. Hoffman) and concluded: "Despite reservations, the collection of cases that Sternglass presents would seem to indicate that there is a relationship between fallout and infant mortality of the general nature he suggests."[16]

Dr Sternglass, a past Chairman of the Federation of American Scientists, Pittsburgh Branch, worked for Westinghouse from 1952 until 1967. He is now Professor of Radiation Physics and Director of Radiological Physics in the Department of Radiology, University of Pittsburgh. While with Westinghouse, he was Advisory Physicist reporting to the Vice-President and Director of Westinghouse Research Laboratories, where he did much detailed research in nuclear physics. Thus the oft-heard jibe that anti-nuke scientists are those who have failed to get to the top can hardly be applied to him.

Nor can it be applied to Dr John Gofman, who in 1963 was invited by Glenn Seaborg, then Chairman of the AEC, to undertake a long-range study of the biological effects of radiation and report on the dangers to man and other species that might result from a nuclear power programme.

Six years later, in October 1969, Dr Gofman stood up at a routine meeting of the Institute of Electrical and Electronic Engineers (IEEE) and announced that if America's nuclear power programme expanded according to plan, low-level radiation would eventually cause an extra 24,000 cancer cases per annum.

Had the speaker not been Dr Gofman, this apparently crazy prediction would certainly have been ignored. But it was impossible to ignore a man internationally renowned as one of the foremost research scientists of his generation. Dr Gofman was then Professor of Medical Physics at the University of California and had won the two highest American awards for his work on heart diseases. He also holds a PhD in nuclear chemistry, as a nuclear chemist has made several original discoveries, and is a co-discoverer of uranium-233. For years he was one of seven associate directors of the Lawrence Radiation Laboratory (LRL), America's largest and best-equipped centre for the design of nuclear weapons. And, during his six-year study for the AEC, he enjoyed an annual operating budget of $2 million and had full access to all government information on US nuclear programmes.

After his IEEE speech, Dr Gofman was invited to testify before the Muskie Committee on Air and Water Pollution; but the AEC forbade him to mention again in public the results of his radiation research. At the LRL he had also been conducting a $250,000 study of the relationship of chromosomes to cancer and had won international acclaim for the initial stages of this project; soon the AEC made it plain that all his funding would be withdrawn and he was forced to resign from his LRL directorship. As a man immensely revered within the independent scientific community, he could at once have resumed his Berkeley researches — though with much reduced funding. Instead, he helped to found America's anti-nuke movement, which owes much of its strength and "respectability" to John Gofman's stature among the world's scientists. His colleagues do not always agree with his findings, but no Gofman finding in the field of radiation biology can be scorned as the vapourings of an econut.

Among Dr Gofman's close collaborators at the LRL was Dr Arthur Tamplin, earlier a RAND corporation space programme celebrity, who willingly followed his leader into the unfunded wilderness. Together, Gofman and Tamplin applied themselves to a detailed critical analysis of the amounts of radioactive effluents permitted to be discharged from nuclear sites. In the US, standards are set for these discharges by the NRC, which is advised by the NCRP, a private government-approved body which examines advice from the ICRP before providing recommendations to the NRC. Both the NCRP and the AEC disagreed with Gofman and Tamplin when they supported the "doubling dose" concept — meaning the dose of radiation which might be expected to double the incidence of any pathological effect. However, the accepted doubling-dose has now been much reduced.

Soon after, Tamplin and Cochrane published a controversial report suggesting that to inhale minute dust particles of plutonium — as little as one-millionth of a gram — might have a serious effect. They argued that the very highly concentrated short-range alpha emission from such a "hot particle" intensely irradiates a microscopic area of lung tissue. Thus the effect cannot be accurately forecast on the basis that it is spread over the whole lung, as present standards assume. Tamplin and Cochrane are adamant that the permissible concentration of plutonium oxide in the air should be reduced by a factor of 115,000.[17] Neither the American nor the British nuclear industry accepts this and Flowers noted (para. 69): "No case has been made in favour of the special carcinogenicity of 'hot particles'. Indeed, there is strong circumstantial evidence to the contrary. The suggestion does not, therefore, provide a reason for dramatically tightening standards beyond those that are now generally accepted . . . Nevertheless, the problem is so important that we would like to see further experiments mounted to confirm the correctness of the current conclusions."

Meanwhile, up to 6,000 curies of plutonium may legally be dumped into the Irish Sea each year and the radioactive emissions from Windscale are detectable off the coast of Norway. The sea around Cap La Hague — where the French have their reprocessing plant — is also troubled, showing radiation levels more than five times higher than normal. No wonder an increasing number of people feel that the "correctness of current conclusions" should be confirmed *before* any further releases of plutonium take place. Once committed to the oceans, this isotope can never be retrieved — except through the food chain, or by inhalation if natural events release particles into the atmosphere. Flowers acknowledges that this could happen (para. 352): "It is known that the estuary contours (around Windscale) are changing with time because of the net landward movement of sediment. Thus in about a century our descendants may be faced with two new exposure pathways for this plutonium. Sediments that are above high-tide level and are no longer wetted may blow about in dry windy weather and possibly form respirable aerosols. A smaller hazard could arise from grass growing on newly reclaimed land, which may support cattle as happens now. They would require careful monitoring to ensure that there is no radiological hazard."

In 1947 it was realised that plutonium accumulates in the marrow among our blood-forming cells. In 1967 J. Vaughan, B. Bleaney and M. Williamson pointed out in the *British Journal of Haematology* that a consequent leukemia risk was possible. But, like Tamplin and Cochrane, they were not taken seriously; a 1975 Medical Research Council report[18] scarcely referred to their study. Yet on 13 January 1975 an article by A. Tucker in the *Guardian* had drawn public attention to the fact that "Figures Show Leukemia Link with Plutonium Workers". As Flowers explained (paras 73 and 74): "Leukemia had developed in several of the workers at Windscale and other plants where plutonium is extracted or fabricated. BNFL and the NRPB denied that any deaths attributable to plutonium had occurred, but we felt that the facts needed to be clearly established and we asked the NRPB to do this and set them out for public examination. In September 1975 the NRPB produced a paper in which a comparison was made between the numbers of observed and expected deaths at Windscale alone, from leukemia and other types of cancer, including those involving the bone marrow. They concluded that the excess of observed deaths was not statistically significant and appeared to regard the matter as closed. Yet the data they reported were limited to observations on men while they were actually employed and were therefore estimated to have omitted at least 50 per cent of the deaths from cancer that would have been found had the men continued to be kept under observation after they had left employment. It is a common experience in industrial medicine to find that observations limited to the period of employment are biased by a deficiency of deaths from cancer and chronic diseases and it is difficult to see why it has not been possible to carry out a proper study of all radiation workers, whether or not they have ceased employment, as has been commonplace in other industries. It may be noted that any leukemias from plutonium would be expected to have developed after a longer time than usual because of the slow rate of translocation of plutonium to the bone marrow."

In the US in 1964, Dr Thomas Mancuso was asked by the Energy Research and Development Administration (ERDA) to conduct a study to discover what, if any, biological changes had been induced by low-level radiation among the nuclear workers at Hanford, Washington, and Oak Ridge, Tennessee — America's oldest and largest atomic establishments. Dr Mancuso analysed one million files and the death-certificates of 3,710 former workers, but by 1974 had not completed his research. The AEC then urged him to publish his finding to refute Dr Samuel Milham, of the Washington State Health Department, who had just concluded an independent study and reported a high rate of cancer among former employees at Hanford — where plutonium has been manufactured since 1944. When Dr Mancuso refused to publish, claiming that he needed more time, the AEC told him that in July 1977 his funding would be withdrawn and he would be required to hand his data to the Oak Ridge government-run laboratories. Immediately he sent an SOS to Dr Alice Stewart and her colleague, Dr George Kneale, a biostatistician. Both hurriedly arranged sabbatical leave from Oxford and all three worked together on the Hanford material, temporarily setting aside the Oak Ridge figures. In 1977 they confirmed Dr Milham's results: an increase in all cancers, especially those of the lymph and bone marrow. At doses far below the accepted workers' limits, there was a doubling of the risk of bone-marrow cancer. Dr Mancuso, a physician and Professor of Public Health at the University of Pittsburgh, concluded that the dose required to double a person's cancer-chances is less than half the internationally accepted limit.[19] His views were supported by Dr Radford, Chairman of the influential Biological Effects of Ionizing Radiation (BEIR) Committee, who said — at Congressional hearings on the Mancuso study — that the present levels are over ten times too high and over a forty-year period double the chances of workers getting cancer.[20]

Dr Thomas Najarian, a young haematologist, is also studying nuclear industry workers. By chance he observed a high incidence of blood-related diseases among youngish employees in the Portsmouth US Naval Shipyard, where nuclear submarines are maintained. At first the US Navy refused him access to its medical records, but a study of workers' death certificates showed that 33 per cent of shipyard workers had died of cancer (twice the national average). Those who had been exposed to accidental leakages of radiation had leukemia rates four-and-a-half times the national average, and amongst older workers who had been exposed to leakages 60 per cent had died of cancer. After Dr Najarian had published his initial findings, the Navy agreed to a full-scale study.[21]

Pro-nukes have also tried to obstruct the work of Dr Rosalie Bertell, a Roman Catholic nun who is Assistant Research Professor at the State University of New York in Buffalo and a bio-statistician at the Roswell Park Memorial Cancer Research Institute. In 1978 she explained:

I became involved in the nuclear power controversy when I experienced intimidation and resistance from a utility company when trying to inform the public of my research findings on the environmental causes of leukemia ... My studies are based on actual measurements of radiation-related human health effects at the levels of exposure routinely given to workers and to the general public from the nuclear industry, nuclear weapons production and testing. They are not projections or forecasts, based on the very high-level radiation exposures at Hiroshima and Nagasaki, such as the nuclear industry and the ICRP are using. In the business world, a forecast is set aside as soon as direct audit information is available! ... These research findings have been presented at professional meetings, were subject to peer review and are published in recognised scientific journals. Dr Bross's work on the genetic effects of radiation has been chosen as the outstanding cancer research of 1977. In May 1976 we requested formal hearings before the US NRC ... A preliminary meeting was finally held after two years' wait ... This was non-official and could result in no regulatory action ... Isn't it strange that Dr Sidney Marks, after criticising Dr Mancuso for the Department of Energy, went out to Batelle Northwest and accepted a Department of Energy contract to re-analyse Mancuso's Hanford data? Isn't it strange that Dr Marks was chosen to represent the US at the IAEA Conference on the hazards of low-level radiation at Munich, March of this year? The US representative had to be authorised by the Department of Energy. No scientist dealing with first-hand data on the hazards of low-level radiation, and speaking out against needless human misery and death, was authorised to attend. Dr Bross, my immediate supervisor at Roswell, did make a request to attend.[22]

Dr Bross soon after suffered the same fate as Gofman, Tamplin, Mancuso and Milham: his funding was cut off and he was subjected to an extraordinary barrage of *personal* attack. Such unscientific behaviour is also observable in Britain. On 27 October 1979, at a conference on low-level radiation at Guy's Hospital, the NRPB representatives made no attempt to provide evidence supporting ICRP recommendations but confined themselves to personal attacks on the "opposition".[23]

From all these dedicated anti-nuke crusaders the diligent seeker after truth must now turn to the pro-nuke spokesmen. These, it has to be admitted, are superficially more plausible. For most people, it is easier to believe that regulated releases of low-level radiation are harmless than it is to believe that one-millionth of a gram of plutonium may cause a fatal cancer. More-over, the average employee is plainly unafraid of such radiation: otherwise he wouldn't be working where he is. I have however met two exceptions, the younger of whom is about to leave the industry. His older colleague reckons it's too late to worry but is dissuading his son from following in father's irradiated foot-steps. The morale of Britain's nuclear workforce was perceptibly lowered by the Windscale Inquiry. Nor was it raised when the widows of two Windscale leukemia victims, who died at the ages of thirty-six and thirty-eight, were offered £67,000 and £28,000, respectively, by BNFL and the AEA.

In a leaflet entitled "The Need for Nuclear Power", the Electricity Council of the UK claims: "More than 200 reactor years of experience have been accumulated without observable injury to anyone." At least they have had the grace to insert the word "observable", no doubt hoping its implications will be lost on the Great British Public.

Various questions about radiation are asked and answered in this leaflet.

"*Q*. What about radioactivity escaping from a nuclear power station? Isn't it dangerous to live near one?

"*A*. The amounts of radioactivity emitted from nuclear power stations are extremely small, and certainly much less than you encounter in other ways. For example, the average background radiation level is about 100 radiation units a year, or 150 if you live in a granite house in Aberdeen. Compared with these figures, the additional radiation received on average by a person living near a nuclear power station varies from less than one-half to less than one-twentieth of the average background radiation level. So you can see that living near a nuclear power station is not very different from moving from one part of the country to another."

This sounds most reassuring, unless one is neurotic enough to worry about both the *accumulative* effects of *any* additional radiation and the amounts received *not* on average by persons living near nukes.

"*Q*. What about the highly radioactive fission products?

"*A*. Perhaps the most important point is that the amounts involved are so small. For example, if a man's total energy requirements for his entire lifetime were met from nuclear energy, then in glassified form the active nuclear waste would literally amount to a handful."

So small is beautiful, after all! But we are not told that quite a few lethal handfuls are produced annually by each reactor: some 500 pounds of plutonium, not to mention the other long-lasting radioactive materials (actinides) produced. This disingenuous equation of *smallness* with *harmlessness* does nothing for the credibility of the Electricity Council.

"*Q*. But what happens to this waste?

"*A*. At present the waste is stored in liquid form in steel-clad tanks at Windscale. Because the amounts involved are so small, there is no great rush to decide what to do with this waste in the long term. It seems likely that it will be glassfied, placed in stainless-steel containers, and stored either deep underground or in the ocean depths."

The effrontery of this attempt to deceive the public is quite breathtaking. In September 1976 Flowers stated (paras 365, 367 and 391): "The amount of solid waste, all of it significantly contaminated with plutonium, being stored currently at Windscale is some 12,000 cubic metres, and within this volume there is a little under half a tonne of plutonium. There is also a growing amount of americium . . . It may prove difficult in the future to retrieve and package this material for ultimate disposal. In particular, the presence of considerable quantities of water with high radioactivity content represents a hazard . . . Neither the AEA nor

BNFL in their submissions to us gave any indication that they regarded the search for a means of final disposal of highly active waste as at all pressing, and it appears they have only recently taken firm steps towards seeking solutions. We think that quite inadequate attention has been given to this matter, and we find this the more surprising in view of the large nuclear programmes that both bodies envisage for the coming decades, which would give rise to much greater quantities of waste."

"*Q*. Isn't this waste a threat to future generations? After all, it's lethal for millions of years.

"*A*. No. The major part of the active waste loses its radioactivity within 500 to 1,000 years. A very small proportion of highly active elements have longer lives, perhaps a million years or more. If this residual activity reached human environment it would be only a fraction of natural radiation levels. To get this question into perspective, look at other forms of poison, such as arsenic, or mercury, or cyanide. They never decay. They last for ever."

As this book must by now have made plain, no one can foretell the results of such "residual activity" working its way back into the human environment. And it is tendentious to refer to high-level waste as a "poison", like arsenic, mercury or cyanide. True, these lethal substances never decay; yet because they are not radioactive they present quite different and much less intractable disposal problems.

Another leaflet in the Nuclear Power Information Group pack (which was widely advertised in the British press during 1980) comes from the UKAEA and is entitled "How Safe is Nuclear Power?" It informs us that: "All radiation discharges to the environment are subject to authorisation by appropriate Departments or Ministries, while conditions inside nuclear establishments are, in addition, subject to the provisions of the Health and Safety at Work Act . . . All nuclear installations are subject to periodic independent inspections, waste discharges are checked and the levels of radioactivity in the environment are monitored."

Readers who believed all this must have been disillusioned by certain items in their newspapers on 1 August 1980. The *Guardian* reported: "In the strongest attack ever made on a public utility, the Health and Safety Executive yesterday condemned the safety standards and professional judgements of British Nuclear Fuels Ltd."

The main *Financial Times* headlines read: "Windscale Safety Rules Neglected, Management Admits." And David Fishlock, the Science Editor, reported:

British Nuclear Fuels Ltd last night admitted "errors of judgement by management and departures from safety standards" in connection with a leak of deadly quantities of highly radioactive acid at its Windscale factory in Cumbria. The Company, which was strongly censured yesterday in a report from the Government's Nuclear Inspectorate for its lack of safety-consciousness and sound professional judgement, says its senior management "accepts full responsibility for the lapses". The Nuclear Inspectorate estimates the leak lasted eight years and amounted to about nine cubic metres of acid totalling more than 100,000 curies of radioactivity. The report, signed by Mr Ronald Gausden, chief nuclear inspector, concludes that BNFL, a wholly owned subsidiary of the UKAEA, failed to comply with several conditions of its site licence for Windscale. But the company had not breached the Health and Safety at Work Act. Radioactive contamination caused by the leak is confined about fifteen feet below the surface, close to the B701 building from which it occurred. The report says it "has not, so far, presented any hazard to workers or members of the public and is not likely to do so in view of the remedial action being carried out". The B701 building was used from 1953 to 1958 but has not been used since. Unknown to Windscale management, highly radioactive fluid was periodically channelled into B701. Tanks overflowed and eventually leaked into the ground. BNFL management told the nuclear inspectors that before the discovery of the leak it "knew of no reason to treat the liquors in this building as radioactive, nor did they think it necessary to check whether the liquids arising in the plant were radioactive". Consequently the Company did not think that the nuclear site licence applied to B701. Mr Gausden makes it plain that he is worried by such statements. Maintenance of safe operational systems "even if well conceived, demands a safety consciousness by management together with sound professional judgement. This was lacking and was the main cause of the incident." Mr Norman Lamont, Under-Secretary for Energy, said yesterday the Government had told BNFL such an incident must not recur. BNFL described the report as a "fair, accurate and comprehensive review of the circumstances". The conclusions were broadly the same as those of its own internal board of enquiry, which reported last autumn. Some of the underlying management problems had already been identified before the leak was discovered, and the Windscale management had taken action. The Health and Safety Executive, to which the chief nuclear inspector reports, said it had seriously considered the possibility of prosecuting the Company. Mr John Locke, director of the Health and Safety Executive, said the legal view was that such a prosecution could not succeed, for several reasons. One was that the accident had not hurt anyone, and that the Company could show that there were back-up safety-systems which would have prevented anyone being hurt by the leak. Mr Locke's department has carried out its own "safety audit" of Windscale, along lines it has used on several other UK companies.

Reading this account of a remarkable exercise in public relations, I recalled some of the statements made almost four years previously by Dr D. G. Avery, Deputy Managing Director BNFL, at the Public Hearing on the Commercial Fast Breeder Reactor held in London in December 1976. Graham Searle, Director, Earth Resources Research Ltd, cross-examined Dr Avery, in particular about Windscale's infamous leaking silo.

"*Searle*: Isn't the point at issue not so much the actual danger of the water that has escaped from the leak but the fact that there was a leak and that there may be leaks in the future? Could there not be leaks of a more serious character?

"*Avery*: There are no absolutes. But the leak about which there has been so much publicity was from the first of these silos that we made. Subsequently their designs have been improved so that, for example, they have been given double walls and should the first wall leak then there are detecting instruments in the space between so that a leak would be detected even before it got to the second wall. This is symptomatic of what's been going on over the last twenty years; a steady progress in safety devices, with multiple containments, so that we are much more confident now than we had any right to be twenty years ago."[24]

Even as Dr Avery spoke, in 1976, the B701 building was steadily leaking "deadly quantities of highly radioactive acid" into the earth — and would continue to do so for the next few years, undetected. No number of safety devices are proof against human frailty. Flowers commented (para. 368): "We have no reason to think the operations at Windscale are not conducted with every attention to safety. Nevertheless, it is important at such a plant that the highest standards of general housekeeping should be employed and we feel bound to say that we did not gain the impression that this was so at the time of our visit. (November 1974)." In November 1974, B701 had already been leaking for some three years; it is of the nature of the problem that the most alarming threats can go unnoticed when experts tour a nuclear installation. This particular paragraph is a good example of the Commission's tendency to fall between the pro- and anti-nuke stools. Some of its members were appalled by Windscale's slovenly management, yet they balked at condemning it outright, without that craven double negative to soften the blow of what followed.

According to the AEA: "All nuclear installations are subject to periodic independent inspections, waste discharges are checked and the levels of radioactivity in the environment are monitored." Why then, at some stage since 1958, did the government's Nuclear Installations Inspectorate (NII) not notice the odd goings-on within and beneath B701? In an inside article on the leak, David Fishlock quoted Mr Con Allday, BNFL's chief executive: "People forgot it was there, and they shouldn't have." Mr Fishlock notes: "The oversight strengthens the hand of those who claim that nuclear waste is too dangerous and long-lived to store permanently on land."

The "oversight" (a mildish term in the context, some will think) also strengthens the hand of those who share certain doubts expressed by Flowers (paras 280, 281, and 282): "The NII is a body of some ninety engineers and scientists which, since 1975, has been part of the Health and Safety Executive (HSE). The Executive reports to the Health and Safety Commission, which formally issues site licences for the whole of Great Britain ... Nuclear site licences ... specify in considerable detail the conditions under which the plant may be operated ... The operator can be ordered to close down a reactor or reduce its rating at any time. These reserve powers are occasionally used ... Although the NII can prosecute operators for infractions, they have not needed to do so ... The NII thus have very substantial powers to control how nuclear reactors (and other installations such as those used for uranium enrichment, fuel reprocessing and radio-isotope production) are designed, built and operated ... We do not doubt the technical competence of the Inspectorate or the thoroughness with which they carry out their work. However, the discussions we have had with several authorities on reactor safety have left us with some doubts about whether the criteria adopted by the NII in establishing reactor safety are soundly based and whether their functions are correctly defined. We have not investigated these matters deeply and we express our views ... with the object of assisting their further consideration by the authorities concerned."

On 15 May 1980 *The Times* reported: "The NII is 20 per cent understaffed and dangerously demoralised; a recent HSE recruiting drive has not attracted even one applicant." Yet during 1980 the NII's work-load was much increased by the indispositions of ageing Magnox reactors and by the urgent need to evaluate PWR design for an impatient government. Also during 1980, a confidential Ministry of Defence report[25] commented that the Aldermaston nuclear weapons factory had only 41 per cent of the health physicists needed to ensure worker safety.

Flowers expressed concern (para. 292) about "the need for a source of independent, expert advice to the Government on technical matters that are relevant to policy decisions on major and hazardous technological developments . . . So far as nuclear matters are concerned, it appears that at present the main source of authoritative advice is the AEA, a body that is inherently committed to nuclear development."

This point was developed by Sir Kelvin Spencer in a letter to the Business Editor of *The Times* (12 November 1980):

The need for nuclear power has been a conviction of a succession of Cabinets since the UKAEA was set up in the 1950s when I was Chief Scientist at the Ministry of Fuel and Power. Cabinet ministers haven't time to study complex issues. They must rely on advice from what they hope are competent and unbiased sources. In the nuclear field there are four main sources: civil servants, the staff of the UKAEA, industry and the financial world. Civil servants with any deep knowledge of nuclear complexities left government service for the UKAEA or industry long ago . . . Hence the advice civil servants give to ministers is what they themselves get from the other three sources. It needs no crystal ball to see what advice the UKAEA will give. Advice from industry will be much the same. A firm that gets contracts for anything to do with nuclear power gets paid for good work, and paid for putting right work not so good. Their interest is to keep the nuclear bandwagon rolling. International finance is deeply committed to the mining and marketing of uranium ore. Massive investments have been made in the belief that there will be an expanding market for it. That depends on there being a continuing programme of nuclear power stations. Advice from the City therefore can hardly be relied on for impartiality. But government is faced with growing opposition from the general public. How better could this be countered than . . . by the myth of *need*. The alleged need is given plausibility by playing down the alternatives — the so-called renewables . . . Public funds for developing these are administered by the Energy Technology Support Unit. This is located at Harwell, the centre of nuclear technology. This tilts the scales against "renewables" from the outset . . .

So it's all in the family, which may be why the HSE decided against prosecuting BNFL. Windscale's B701 spectacular must have infuriated the government, which had recently spent some £600,000 on the Electricity Council's propaganda campaign to burnish the nuke image. Inevitably this "deadly leakage" made people wonder how many similar "oversights" there have been, as yet undiscovered or unacknowledged. Yet to have prosecuted BNFL, and thus probed the depths (in every sense) at Windscale, might have permanently tarnished that precious nuke image. Certainly a lot of awkward questions would have been asked. For instance, what exactly were the "back-up safety systems which would have prevented anyone being harmed by the leak"? As no one was aware of this hazard for eight years, the layman finds it hard to imagine them. The supposedly disused B701 was assumed to be harmless, so what — in any case — were such safety systems backing up? Again, how can BNFL be sure that those nine cubic metres of highly radioactive acid are "confined" some fifteen feet below the surface? What is confining them? If highly radioactive waste is so easily confined, why are governments spending millions on waste-disposal r & d? And how can BNFL be so confident that those 100,000 curies of radioactivity, astray in the Cumbrian soil, have endangered neither workers nor the public? Given the present state of our knowledge about radiation, this claim must be regarded with some caution.

The nuclear industry is very dependent on a mixture of bluff and public ignorance: and as the latter diminishes the former becomes less effective. This process was well illustrated by the West's reaction to Russia's Urals tragedy.

In late 1957 or early 1958 a massive amount of high-level radioactive waste exploded on an atomic weapons site in the Chelyabinsk region. According to CIA documents released in December 1976: "At the time of the accident, highly poisonous radioactive cobalt, barium, cesium and strontium had been in the storage tanks about ten years" — and, it seems, had been inadequately monitored.[26] The most reliable estimates reckon that primary radioactive contamination covered between 800 and

1200 square miles. Nobody knows how many died — the region's population was approximately 200,000 — but entire villages had to be evacuated and levelled. The remaining plant and animal life was so heavily irradiated that major mutations are expected. For years drivers had to speed across the area, in sealed vehicles. By now the industrial towns have been decontaminated and covered with new asphalt, but the countryside and villages remain deserted.

Had the CIA at once revealed this accident the Russians would have been gravely embarrassed — but so would America and Britain. The 1957 Windscale fire was still fresh in British memories and public hostility to atmospheric bomb-testing was at its height. When it comes to the crunch — or the bang — nuclear hawks flock together, whatever their ideological differences.

The Urals disaster was publicised in 1976 by Dr Zhores A. Medvedev, a dissident Russian scientist now working with the National Institute for Medical Research in London. The CIA then reluctantly admitted that they had been aware of the catastrophe, but — contradicting themselves — claimed that it was no more than a major reactor accident, the results of which were soon cleaned up. Britain's pro-nukes were equally disingenuous. In a Press Association interview, published in *The Times* on 8 November 1976, Sir John Hill, then Chairman of the AEA, dismissed Dr Medvedev's account of the disaster as "rubbish", "pure science-fiction" and "a figment of the imagination". He insisted that high-level waste "could not give that kind of explosion, nuclear or thermal".

Subsequently, Dr Medvedev, a specialist on radioactive isotopes, did a vast amount of research on the relevant Soviet scientific papers and journals, which are freely available to any Russian-speaker. These provided proof of his original assertion, which he published in book form in 1979.[27] When the AEA were asked for their reaction to this book they sourly replied: "No comment." Too much evidence had been accumulated for any more bluffing about "pure science fiction".

On 8 November 1980, a day-long symposium was held in London, at the Royal College of Medicine, to consider the medical consequences of a nuclear disaster — either civilian or military. I eagerly attended, armed with notebook and pencil, and discovered that some of my friends in the British nuclear industry had misinformed me about the possible wartime role of nuclear power stations. (No doubt they spoke out of ignorance, as this aspect of the nuclear scene has only recently been studied.) When questioned about the likely effects of a conventional weapons attack on a nuclear power station, most of them had assured me that the reactors, being contained in vessels with immensely thick concrete walls, would survive. However, it has now been pointed out[28] that the latest thing in precision-guided conventional weapons could either cause a meltdown or rupture both the containment and pressure vessels, thus releasing and dispersing most of the fission products.

From the attacker's point of view, this would neatly dispose of three condor-sized birds with one stone. Vast numbers of the enemy population would be irradiated, all for the price of one conventional weapon. The victim nation's electricity supply would be disrupted and its industrial activities hampered. And, because reactors represent such an enormous concentration of capital investment, the national economy would be grievously damaged. For a nukeless enemy, or an enemy who wished not to be the first to use nukes, these targets could present an irresistible temptation. Britain is not yet sufficiently dependent on reactors to make them prime targets *for motives of industrial disruption*. But if all Mrs Thatcher's fissile dreams come true, the UK will present an even greater temptation than at present to paralysing attacks from nukeless enemies.

It has long been agreed that it makes no sense to worry about a nuclear bomb being dropped on a nuclear reactor, since the bomb itself would cause such hideous devastation. But now Steve Fetter and Kosta Tsipis of the MIT[29] have explained that an attack on a nuclear power station with a 1000 kiloton nuke would vaporise the whole reactor and draw at least 3 per cent of its radioactive contents into the fireball, thus *altering the nature* of the subsequent fallout. This addition to the isotopic composition of the fireball would considerably reduce whatever chance a nation might have of eventual recovery.

In a nuclear bomb, at the instant of explosion, only the primary fission fragments are present and most of those are short-lived isotopes which later decay into isotopes of longer half-life. In a nuclear reactor, those longer-lived isotopes are steadily built up, being continuously produced as the decay of short-lived isotopes. Therefore the proportion of fission products of long half-life is much greater in a reactor than in a bomb — and so the overall decay of the radioactivity is much slower.

At the symposium, a graph was shown comparing the decay of radioactivity for a one-megaton bomb with that for a one gigawatt reactor in which the fuel elements had remained for three years. The decay of the activity in the reactor was reckoned from the moment of its shutdown and initially the total activity in the reactor was about 100 times lower than that from the bomb; but after some four days they became equal. From then on the reactor's activity was greater and after one year it was still as intense as the bomb activity after one month. All of which means that a nuked reactor would render large areas uninhabitable for far longer than "pure" bomb fallout could do. Moreover, this particular study assumed that only 3 per cent of the radioactive content of the reactor was sucked into the fireball — an extremely conservative estimate, in the opinion of many experts.

Other chilling slides were shown to illustrate the possible effects of nuking the Wylfa and Hinkley Point reactors. If these were attacked when the wind was from the west, the Midlands would be heavily contaminated. And presumably Ireland would be at the receiving end if the wind were from the east. There is a certain irony about the extent to which the long-term effects of these ultra-sophisticated precision weapons would be determined by fickle air-currents.

A year after a one-megaton nuke had been dropped on the PWR proposed for Sizewell, the uninhabitable area would be ten times larger than if the same bomb had *not* damaged a reactor. It would extend to the South Coast, including Greater London, and almost certainly some people would be forced to occupy contaminated areas which gave wholebody doses of 10, 50 or more rem/year. Most probably these levels would have to be tolerated by many survivors, though they could cause serious illness and/or death.

The Fetter and Tsipis study also considered the effects of a nuclear attack on a high-level waste storage tank and analysed the store at West Valley, New York State, which contains the same range of major isotopes as the Windscale tanks. It is worth remembering that there is no plan to provide permanent alternative storage for these wastes before the year 2000. And if a nuke dispersed the radioactivity of just *one* Windscale tank, the long-term damage would be even more severe than if a reactor were nuked because the total cumulative radioactivity of these wastes would be released.

In the context of attacks on nuclear installations, Western Europe is the most vulnerable area. Only one nuked West German reactor would render most of that country uninhabitable for at least a month, even if people accepted cumulative doses much higher than 2 rem. At a time when "limited" nuclear war is being described as "part of the Western flexible response", the possibility of such attacks must be faced. Yet in Britain no Civil Defence preparations known to the public take this possibility into account and the 1980 HMSO publication, *Nuclear Weapons*, ignores it.

Towards the end of 1980, several doctors belonging to the American "Physicians for Social Responsibility" organised a symposium, with military and scientific experts, to re-examine the medical consequences of nuclear war. They concluded that it was essential to inform the international medical profession quickly of those consequences and to appeal to them to educate the public — even at the risk of appearing alarmist. The almost total silence of physicians on this topic encourages the notion that somehow populations can survive a nuclear war as they have survived conventional wars. Millions vaguely assume that after a nuclear war people would still have food and shelter, a normal environment, secure social structures and medical problems no different in type and amount to those that had existed before the bombings. Here ignorance is a perilous form of bliss. The longer we continue to live in this cloud-cuckoo-land, the more likely is nuclear war. As the "Physicians for Social Responsibility" have pointed out: "From the perspective of public health, nuclear war would create the most severe public health problems in all of history, dwarfing major plagues and epidemics in the numbers of dead, injured and diseased." We need to be told repeatedly, in simple language, about the burns, trauma, radiation sickness, leukemias, solid tissue tumours and birth defects that would result from even the most "limited" nuclear war — afflictions which could not possibly be treated in a post-nuclear-war society that had disintegrated under the stress of unprecedented devastation, terror and despair. In any First World country, the medical profession could cope more or less adequately with the *immediate* effects of a major reactor disaster. But they could not even begin to cope with the results of a nuclear bombardment.

6

What is MUF and Where is It?

The sad truth is we've opened a Pandora's Box and there is no way to return to a world safe from nuclear weapons, even nuclear weapons in the hands of terrorists.

DAVID ROSENBAUM, 1975[1]

Nuclear generators are partly a façade for weapons research, partly a source of electricity needed to power a sophisticated weapons technology and partly a weapon in their own right in the international economic war in which we are presently engaged.

ROSALIE BERTELL, 1978[2]

As Dr Helen Caldicott has reminded us, great power can attract unstable types. Rumour has it that some weeks before Richard Nixon resigned, worried officials saw to it that he no longer had access to the nuclear buttons: and if this is not true it should be. However, Russia and America have a more than even chance of being ruled by men who are not button-happy; therefore many feel that proliferation is a greater threat than superpower conflict. The spread of nuclear reactors allows nuclear weapons to be furtively manufactured by countries with chronically unstable governments or unpredictable leaders. Even the most obdurate pro-nukes have to admit, when challenged, that this "presents problems".

Proliferation is an easily understood hazard, unlike high-level waste disposal, threats to civil liberties and radiation. Everyone can appreciate what the spread of nuclear weapons is likely to entail, though the industry has successfully confused most people about the link between reactors and bombs. Recently a keen-witted man, who has had a distinguished career in Whitehall, wrote to me asserting that "the closing-down of every nuclear power station would not make any difference to the number of atomic bombs in the world". The nuclear industry likes us to think thus, but it disturbed me to find such a man uncritically swallowing their bait. For years proliferation has been widely discussed, so one can only conclude that devout pro-nukes ignore anything which might shake their faith, rather as old-fashioned Roman Catholics avoided "heretical" books written by Protestants and such-like.

The first stage of the proliferation process is well illustrated by Iraq's nuclear enterprise. That country — which in September 1980 proved itself among the most radical and aggressive of Arab states — has two reactors, Talmuz-1 and Talmuz-2, on a site near Baghdad. Talmuz-2, the smaller (0.8 MW), is already completed and has been planned as a research centre for the training of 600 engineers and scientists. It needs only a few kilograms of enriched uranium to go critical and was loaded at the beginning of August 1980. Talmuz-1 is a 70 MW reactor, which might have gone critical at the end of 1981 had Israel not intervened.

In 1975 France agreed to provide Iraq with these two research reactors (worth £140 million) and with 93 per cent enriched uranium (suitable for bomb-making) at a rate of twelve kilograms every six months during Talmuz-1's first three years of operation. In April 1979 a vital component of Iraq's nuclear order was sabotaged in its French construction yard. At once France offered to provide instead specially adapted reactors requiring only 8 per cent enriched uranium (unsuitable for bomb-making). But the Iraqis indignantly rejected this suggestion and in these hard oil-times France's not to reason why ... America, Britain and Israel then put tremendous pressure on France to break her 1975 contract, but they were ignored. In June 1980, Dr Yahia El-Mashad, an Egyptian national but one of Iraq's leading nuclear scientists — trained in Russia — was murdered in his Paris hotel room. He had travelled to France to meet officials of the government Atomic Energy Commissariat and his death was calculated to delay Iraq's progress towards nuclear statehood.

Iraq has signed the Non-Proliferation Treaty (NPT) and is therefore pledged not to manufacture nuclear weapons — a pledge unlikely to reassure any moderately shrewd five-year-old. The Iraqis insist that they have no intention of *stockpiling* nuclear weapons: the unacknowledged object of the Talmuz exercise is to provide them with the technology and raw materials to make a nuke quickie in an emergency. Nuke quickies are easily made, with reactor-grade materials. In September 1977 the Americans exploded a bomb filled with "normal reactor quality plutonium" at their Nevada testing-ground. It was a crude thing — the nuke equivalent of a stone axe — but if exploded over a city it could have killed 100,000 people.

Non-nuclear states are rapidly acquiring the knack of going nuclear without becoming dependent on any one country. Again, Iraq provides a good illustration. In May 1979, Brazil agreed to swap nuclear fuel, training and uranium-prospecting help for oil. "Informed sources" referred also to a secret clause promising that Brazil will trade plutonium, once its own West-German-supplied enrichment plant has made this possible — perhaps by 1985. In March 1980, Portugal eagerly agreed to trade an unspecified amount of processed uranium ore. And, most sinister of all, in 1978 Italy agreed to supply £25 million worth of equipment, including a "hot cell" for handling, by remote control, heavily irradiated materials fresh from the reactor core. With its assistance, enough plutonium can be extracted from spent fuel to make a "reactor bomb".[3]

Oil-merchants can't be bossed about. Iraq refused to enter into a formal France-Iraq-International Atomic Energy Agency (IAEA) agreement, though an exchange of letters published in September 1979 revealed that Baghdad had consented to IAEA checks on the destination and use of French-supplied enriched uranium. But none of this matters much since the IAEA is essentially an industry tool. It was established by international statute in 1957 and one of its statutory objectives is to "accelerate and enlarge the contribution of atomic energy to peace, health and prosperity around the world". As Brian Johnson has written: "It represented from the start a compromise between US hopes that foreign nuclear facilities could be adequately safeguarded by internationally agreed safeguards, and the political sensitivities and aspirations of the member states of the UN."[4] Like America's AEC and NRC, and Britain's AEA, the IAEA is dedicated to the promotion of nuclear power. However, it seems to be less unscrupulous in its methods than these other agencies and for political reasons it may be assumed to be genuinely anxious to prevent proliferation in the Arab world. Yet its powers are eunuch-like; it can go so far but no further. If a country resents its supervision it must run home, apologising for the intrusion. To quote Brian Johnson again: "We must here recall the negotiated and secret nature of IAEA's safeguards agreements; the aversion of most countries to even the present low level of physical surveillance; the immense technical difficulties of accounting adequately for all material in a commercial-scale reprocessing plant, and finally, the admitted unsafeguardability of stored reactor-grade plutonium given determination by the supposedly safeguarded state to make a bomb. Given these realities, claims that even an IAEA safeguards system that was strengthened to the limits of political realism could handle such facilities just are not credible. The only answer therefore must be to prevent the export and construction of such facilities in the first place: an answer which lies for some time to come in the hands of the major nuclear exporting states."[5]

During 1973–4 the IAEA prepared a *Market Survey for Nuclear Power in Developing Countries,* at the suggestion of Dwight J. Porter, a former Hollywood promoter who was then US Ambassador to the IAEA. One could base a tragi-comedy on this bizarre report. Among its multiple aberrations was the advice that Bolivia, El Salvador, Guatemala, Panama and Uganda should each build one or two 150 MW reactors. It recommended Bangladesh as a market for *ten* reactors during the '80s (never mind all those cyclones and tidal waves) and Pakistan was urged to build *twenty-four* nuclear power stations before the year 2000.

This cockeyed survey was based on the assumption that the capital costs of reactor construction would be blissfully low in "developing" countries where peons gladly work for a pittance. According to the IAEA fantasy, a Pakistani welder would do half as much work as a Canadian welder for 10 per cent of the wages. Incredibly, this high-powered study group *did not know* that a reactor cobbled together by Pakistani welders would be about as safe as a nuclear submarine commanded by me. When some realistic non-expert had rubbed their noses in this fundamental fact of Third World life they hurriedly did their sums again. Having allowed for the retraining of local workforces, the need for years of close supervision by foreign technicians, the provision of houses, schools and hardship pay for said technicians and their families — and so on — the original estimates were proved too low by a factor of 2 to 2.5 times. And this is the Agency upon which we must rely to prevent proliferation.

Pakistan is of course going ahead with the sort of nuclear programme that suits the Pakistani government, disregarding both the advice and the admonitions of the IAEA. In June 1979 the French Ambassador to Pakistan, and a colleague, were violently assaulted as they approached the hugely expensive gas centrifuge uranium enrichment plant being built some twenty miles east of Islamabad. (Pakistan's only operating nuclear power plant, a Candu heavy water reactor, does not use enriched uranium.) Three days later Chris Sherwell of the *Financial Times* was beaten up, and then arrested, by six muscular plainclothes men who happened to notice him outside the Islamabad home of Dr Abdul Qadir Khan, a Co-Director of the Engineering Research Laboratory at Islamabad Airport.

In the House of Commons on 20 December 1979, Mr Tom Dalyell, a Labour backbencher, claimed that the same Dr Khan, while a *bona fide* nuclear research worker at the Joint Centrifuge Project in Almelo, Holland, was allowed to obtain bomb-making secrets as a result of "mind-boggling inefficiency or naïvety". Mr Norman Lamont, Energy Under-Secretary, agreed with Mr Dalyell that an Islamic nuke is not a good idea. Then he added, inanely: "The Pakistani authorities have given assurances about the peaceful nature of their programme." This is the sort of remark that devalues parliamentary democracy — an important point, but one we have no time to ponder.

A few months earlier, the *Financial Times* had reported that two former Pakistani army officers, now working for the Pakistani Atomic Energy Commission, were sent by Dr Khan to the north-London suburb of Colindale to buy equipment and components for the enrichment plant. There they found themselves in the "front" offices of a nameless English firm. And — so convoluted is the nuke scene — some of their purchases had been made in the little Irish town of Ennis, Co. Clare.

In December 1979, Customs officials at London Airport held up a consignment of goods from Ennis Precision Products Ltd, addressed to an English firm, and found lots of tiny objects known as "expansion joints". According to their experts, these unlikely products of Co. Clare were being sold illegally to Pakistan for her nuclear bomb programme. Mr Eddie Richardson, Managing Director of the Ennis company, later explained that he was completely unaware of the nuke connection. The expansion joint was earning £50,000 a year for his little firm, which has an annual turn-over of about £300,000, and four specially trained men, out of a work-force of thirty-six, were engaged full-time on its manufacture. In an interview with the *Irish Times,* Mr Richardson said that his contract had come from an English firm and "money would appear to be no obstacle, all the client wanted was a good product . . . Several firms in England failed to produce a product that fully satisfied their requirements. We felt proud that the quality of our product was so highly rated." Reading between these lines, one deduces that the exporters of certain "components" have to choose their manufacturers with an eye to more than quality: an innocent Irish firm in the heart of Co. Clare no doubt seemed ideal — one that also makes components for Sweden, Sperry Vickers and several Middle Eastern countries.

Proliferation breeds melodrama. Secrecy is essential, both to elude the detection of "fissile" crimes and because uncertainty has a high value should a country decide to play nuclear bluff.

In 1957, post-Suez, Israel felt it was time to take the nukeward road. Helped at first mainly by France (the two countries then shared Egypt as an enemy), and by a steady flow of highly qualified immigrant scientists, she built a 24 MW research reactor at Dimona in the Negev desert. For three years the Israeli public remained unaware of its existence — until Christian Herter, US Secretary of State, challenged Ben Gurion about rumours that Israel was planning to produce plutonium. Ben Gurion indignantly protested that Dimona would serve only the needs of agriculture, industry and health.

Sixteen years later, in June 1976, the West German semi-official military monthly *Wehrtechnik* reported: "What until now had often been regarded as a pure speculation turned out to be a hard fact: Israel possesses an atomic bomb, more precisely, thirteen bombs, each of them having an explosive capacity of 20 KT, which is equivalent to one of the bombs dropped on Hiroshima or

Nagasaki. These bombs can be delivered to the target by the Israeli Kfir and Phantom fighters which have been specially equipped for this purpose."

Wehrtechnik could not afford to be shocked, since West Germany had helped enthusiastically in the setting-up of the Israeli nuclear programme: notably by financing such installations as the 6-MV-Tandem-van-de-Graaf accelerator, worth DM 6 million, which made possible Israel's department of experimental nuclear physics. But elsewhere great anxiety was expressed about what might be going on at Dimona, especially after Israel had repeatedly refused to consider any plan for a nuke-free Middle East.

America was particularly worried, at least outwardly, and put such pressure on Israel that eventually she had to agree to American scientists inspecting Dimona: where apparently they found nothing suspicious. Yet by 1966 (according to *Wehrtechnik*), Moshe Dayan, then Minister of Defence, had ordered a nuke to be made — in defiance of Israel's National Security Council, which had vetoed bomb-making. And when Premier Eshkol discovered Dayan's insubordination he could only accept his minister's argument: "Israel had no other choice."

This is a grim example of nuclear autocracy. Given a small reactor, and a team of discreet scientists, any hawk can disregard governmental decisions and hatch a nuke. Who knows what goes on now at Dimona? *Wehrtechnik* has explained: "Dimona is not only guarded by troops, but also has a highly developed electronic system and radar screens, working around the clock. It is strictly forbidden to all planes, including Israeli war-planes, to fly over the area. During the Six-Day-War an Israeli Mirage III went astray in the area. The plane was ruthlessly shot down by an anti-aircraft missile fired by their own people. When in 1973 a Libyan civilian aircraft inadvertently approached the area Israeli fighters tried to force the plane to change course. When this proved to be of no effect the plane was shot down and 108 of the 113 passengers were killed." It's rough out there.

From where did Israel get her raw material for all these bombs? According to Barbara Newman and Howard Kohn,[6] who have investigated Israel's nuke sources both in depth and in width, much of it was American MUF (Material Unaccounted For).

In 1957, the Nuclear Materials and Equipment Corporation (NUMEC) of Pennsylvania began operations. Its President was Zalman Shapiro, who had worked as a sales agent for the Israeli Ministry of Defence. In 1965, AEC inspectors found 194 pounds of enriched uranium missing from NUMEC's inventory. Shapiro, astonished, could only think that it had been buried by mistake in the plant's waste pits. The AEC dug, but recovered no more than sixteen pounds. That prompted them to scrutinise NUMEC's record more closely and this time they found 382 pounds missing. At which point the inspector in charge of the investigation abruptly resigned from the AEC to take up another job — with NUMEC.

When other AEC officials urged the Justice Department to pursue the investigation a cover-up was ordered by President Johnson. An uninhibited Justice Department probe might have reached the fact that the CIA had supplied Israel with nuke technology and materials during the '50s, thus contradicting Washington's official policies. The AEC knew nothing of this cover-up until 1977, by which time they had become the NRC. Then Carl Duckett, a former Deputy-Director of the CIA Science and Technology Division, explained that Johnson had ordered the CIA Director, Richard Helms, to let Israel keep her MUF, unbeknownst to other Federal agencies — or even to the Secretary of State or the Secretary of Defense.

To return to the '60s — the next AEC inspection of NUMEC, in 1967, revealed another 190 pounds of enriched uranium "unaccounted for". The AEC then fined Shapiro $1.1 million, whereupon he sold NUMEC — and Israel had to think again.

During 1968 the Israeli Central Bureau of Intelligence and Security (Mossad) organised several neat hijackings — possibly in collusion with the governments concerned as these extremely newsworthy stories were kept out of the world press for a decade or more. In France, the driver of a 25-ton truck carrying government-owned enriched uranium collapsed when a canister of teargas was fired into his cab. Mossad then smuggled this load more than 2,000 miles to Dimona. The same trick worked in Britain. There the load was low-grade uranium — yellowcake — which could soon be used by Israel, whose scientists had by then almost perfected their enrichment technique. Again, in November 1968 a small German-registered freighter, legally carrying 200 tons of yellowcake from Belgium to Italy, disappeared for several weeks. Eventually she reappeared, empty, in a Turkish port — with a new crew and a new name.

The *Jerusalem Post* of 5 May 1977 supported the Newman-Kohn report. Having dutifully given the official Israeli government story: "Israel obtains fuel for the Dimona reactor by extracting uranium as a by-product of phosphate plants along the Dead Sea", it went on to quote Paul Leventhal, whose statements it obviously found more credible. Mr Leventhal, a former Counsel to the US Senate Government Operations Committee for supervising proliferation legislation, corroborated much of what Newman and Kohn had written. He concluded: "The important thing is not the country that got it, but that nuclear material could be stolen. The 200-ton shipment was enough to cover Israel's nuclear fuel requirements for 20 years. It is assumed that Israel has the materials to make nuclear weapons. That is not a surprise."

On 12 September 1977, soon after the Russian and American spy satellites had spotted South Africa's nuke test site in the Kalahari Desert, *Newsweek* reported: "Some US intelligence analysts concluded that the bomb the South Africans had planned to set off actually had been made in Israel." However this may be, the Israeli and South African Ministries of Defence have been cooperating very closely indeed since 1962 — yet perhaps not as closely as the West Germans and South Africans, whose nuke alliance provides the most shattering of all proliferation sagas. But this is such a complex cliff-hanger that it would be absurd to attempt to condense it here. Instead, I recommend you to read *The Nuclear Axis*, by Zdenek Cervenka and Barbara Rogers (Julian Friedmann Books, 1978). An appendix provides facsimiles of those government documents on which the exposé is based.

The Nuclear Axis makes it plain why South Africa was invited to take part in the talks that led to the formation of the IAEA and why she has always held a privileged position on the Board of Governors — despite her consistent refusal to sign the NPT. Even after the detection of her test site in 1977, *all* Western countries on the Board of Governors voted against an African proposal to deprive her of her seat. But perhaps one shouldn't be unduly surprised by this. Pakistan was unanimously elected to the Board of Governors during the period when every major newspaper was carrying stories of her illicit bomb-making activities.

On 4 December 1974 Israel formally announced her ability to make nuclear weapons whenever it seemed expedient to do so; hence Iraq's refusal to accept feeble little "adapted" reactors that could contribute nothing to nuke-production. The Israelis resolved not to give the Iraquis enough time to bomb-make. On 18 July 1980 the Israeli Minister of Transport, Mr Yitz Haim Landau, announced to the press: "If the Iraquis get the nuclear option from France it will be up to us to see that they are incapable of using it. We mean business. For us it's a matter of life and death." Eleven days later the Foreign Minister, Mr Yitzhak Shamir, said bluntly that Iraq's enriched uranium supply could only lead to another Middle East War.

If nuclear technology is allowed to spread in Africa the proliferation problem will become particularly acute because several countries, including the Central African Empire, Gabon, Niger and Zaire, possess ample supplies of cheap uranium. At present the situation is worrying enough. Zaire began to operate a research reactor in 1959 and since 1961 Egypt has been operating a Russian-supplied reactor which is subject to no international or bilateral safeguards. Libya, too, has imported a research reactor from Russia and enjoys a bilateral agreement with Argentina (which has not signed the NPT) for nuclear cooperation.

Many Third World countries bitterly resent the NPT and several governments have stated forcefully that for the future they are not prepared to accept Western technology in a Black Box. Dr Jorge Sabato, a former member of the Atomic Energy Commission of Argentina, has expressed popular feeling on the point: "Our countries are not accepting any more importation of technology as it was in the past in the petrochemical industry, in iron and steel manufacture, in the gas industry and in many other activities."[7] The Brazilian government has signalled the same message: "It was also necessary to take into account the need to ensure the full transfer to Brazil of the technologies involved in each of the areas of the fuel cycle corresponding to the type of reactor adopted. In other words, it was not acceptable to replace one form of dependence by another."[8]

This attitude is of course perfectly understandable when one recalls the savagery with which so many North American companies exploit South America. It is essential for the nuclear powers to stop proliferation: but this can only be achieved by giving up their own nuclear weapons and industries. It just won't do to wag a finger at the Third World and say: "Now! Now! You mustn't play with this part of the game — it's dangerous for small children. Let us show you the safe bits and then you may play with those." Such an approach is intolerable on the stage of world affairs in the 1980s.

In April 1973, in Brazil's weekly journal *Machete*, Murilo Mello Filho wrote: "For Brazil the bomb is a kind of military imperative, a political necessity . . . The current world situation allows us no lighter or more agreeable choice." On reading this, Washington jumped like a shot rabbit and made frantic efforts to block Westinghouse's future reactor exports to Brazil. (They had supplied the first reactors.) However, the nuclear industry is an octopus-like organism and if one tentacle is paralysed another will at once go into action.

Brazil and Germany have always had close nuclear contacts. In 1953 Admiral Alvaro Alberto, Chairman of the Brazilian Scientific Council, visited Germany and met two physicists, Paul Harteck and Wilhelm Groth, who in 1939 had called the attention of Hitler's War Ministry to the military applications of atomic fission. During the war these men developed the gas centrifuge method of separating uranium-235 for weapons production. When Alberto met them, Germany was still forbidden any kind of nuclear technology. The Germans then signed a secret contract, guaranteeing to provide special training for three Brazilian chemists, and the necessary equipment for setting up uranium enrichment installations in Brazil was ordered by Groth from fourteen German firms. But the Americans won that round; they detected Groth's equipment in a German port, ready for shipment to Brazil, and triumphantly confiscated it.

Twenty-one years later, in 1975, the Germans had their revenge when Brazil negotiated to buy from them, instead of from America, a complete nuclear fuel cycle kit said to be worth £5,000 million. This deal includes uranium prospecting, mining, processing and enrichment, the construction of nuclear power stations and the reprocessing of spent fuel. Negotiations were held up while President Carter unsuccessfully tried to prevent the Germans from selling this off-the-peg nuclear industry; Germany is still forbidden to manufacture nuclear weapons and nobody believes that the Brazil deal has an exclusively commercial inspiration. Nor does any sane person wish to see South America's superpowers armed with nuclear weapons.

In the American journal, *Arms Control Today*, Paul Buchanan has written: "The Argentine nuclear programme will be oriented in the near future according to the political/military objectives of the presently ruling régime and in response to Brazilian military strength and nuclear policy. If regional tensions do not subside, the chances of these prompting an Argentine nuclear explosion are considerable." In 1976 Rear-Admiral Carlo Castro Madero, Director of the Argentine National Atomic Energy Commission, announced that: "We have attained sufficient capacity to manufacture nuclear weapons."[9] As Argentina has an operating nuclear reactor capable of producing 330 pounds of plutonium a year, and a plutonium reprocessing plant, there is no cause to accuse the Rear-Admiral of bluffing.

Proliferation has been stimulated by the fashionable Long Range Cruise Missile (LRCM). Non-nuclear states can legally buy these grisly little gadgets because they are capable of taking either conventional or nuclear warheads. And they are well within the financial range of governments that could never aspire to an ICBM. At 1975 prices, the "cost per unit kill probability" of the Minuteman III Intercontinental Ballistic Missile (ICBM) was about $700,000, but it was reckoned the LRCM unit lethality would be no more than $3,000. LRCMs are also remarkably versatile; they can change and correct their course en route and will take off from a submarine, a ship — even a passenger-carrying Jumbo. And, being generally less than one yard in diameter and six yards long, they are elusive. Their horrible accuracy depends on tiny, lightweight electronic devices and it would be hard — perhaps impossible — for the most astute satellite to discern whether a particular "mini" carried a nuclear or conventional weapon, or whether it was out to kill or simply swanning around for exercise. Yet because this foully efficient device is not a free-falling ballistic missile the American government Jesuitically maintains that it is exempt from SALT-1 limitations.

A multiplicity of nuke organisations lulls a large, though shrinking, section of the public into a sense of false security. The IAEA, the ICRP, the NPT, the SALT agreements: regularly one reads in the newspapers of their expensive conferences, congresses and confabulations, none of which contributes anything much to global safety but all of which issue wordy statements implying that the situation is well in hand.

As we have seen, the NPT, dating from 1970, has the effectiveness of a parasol in a tornado. No country can be compelled to sign it, those who do sign it cannot be compelled to adhere to it and even those who apparently adhere to it need not be deterred by it from accumulating nuke ingredients. Any enterprising government can store its nuclear bits and pieces in different places; then, should the flourishing of a nuke ever seem a good idea, it need only — if it has signed this ludicrous Treaty — invoke Article X, which allows it to withdraw on giving three months' notice "if it decides that extraordinary events, related to the subject matter of this treaty, have jeopardised the supreme interests of its country".

Victor Gilinsky, of the US NRC, has admitted: "Today, after thirty years, we have come almost full circle in our thinking about controls. There are great dangers in building a régime around procedures riddled with exceptions to be taken under political and economic expediencies of the moment. That is how we got into the situation which confronts us today."[10]

The most disarming (no pun intended) of the nuclear acronyms — MUF: Material Unaccounted For — has peculiarly sinister connotations. *A Special Safeguards Study*, prepared for the AEC by Dr David Rosenbaum and a team of security experts, concluded that: "The potential harm to the public from the explosion of an illicitly made nuclear weapon is greater than that from any plausible power plant accident, including one which involves a core meltdown and subsequent breach of containment."[11]

Two other pro-nukes, a law professor and a nuclear physicist, have written: "A nuclear explosion with a yield of one kiloton could destroy a major industrial installation or several office buildings costing hundreds of millions to billions of dollars. The hundreds of thousands of people whose health might be severely damaged by dispersal of plutonium, or the tens of thousands of people who might be killed by a low-yield nuclear explosion in a densely populated area, represent incalculable but immense costs to society."[12] Yet these two men urged an expansion of the nuclear industry, so confident were they that the safeguards could be made sufficiently fail-safe. That of course was way back in 1974: since then, a number of relevant nuke skeletons have come tumbling out of cupboards.

By September 1976, in America alone, 8,000 pounds of plutonium and enriched uranium were "unaccounted for".[13] The publication of these figures naturally caused considerable alarm and despondency, and in 1979 the NRC forsook "MUF" in favour of the vaguer "Inventory Difference" (ID) — which semantic exercise is unlikely to placate an increasingly edgy public.

Nuclear blackmail has to be taken seriously: as it was on 27 October 1970, in the city of Orlando, Florida, when the Mayor got a note demanding $1 million and a safe escort to Cuba, in exchange for not blowing up the city with an H-bomb. An enclosed sketch of the blackmailer's weapon convinced scientists that the threat could be genuine. And the AEC had to acknowledge that ample "special nuclear material" (SNM) had been "diverted" — to use their own genteel expression — for the manufacture of many such bombs. As the Mayor was about to pay out, the blackmailer was found to be a fourteen-year-old high-school student of unusual scientific talent who had designed his sketch-bomb from printed materials publicly available.

Early in 1979 a reputable journalist, Howard Morland, pointed out in an article for *The Progressive* that every element of the nuclear weapons "secret" has long since been made available in encyclopaedias and other readily accessible sources. The information had never — he thought — been drawn together and published as a do-it-yourself manual, yet it was known to numerous college students and could easily be assembled by the scientists of any interested country.

The reaction to this article was strange, in view of the Orlando incident and the many public admissions by officialdom that nuke-making (given a few pounds of "diverted" plutonium) is no problem. On 9 March Mr Morland's article became the first piece of journalism, apart from the Pentagon Papers, to be banned before publication by the US government for "national security" reasons. Affidavits from the Secretary of State, the Secretary of Defense and the Secretary of Energy asserted that the article could help many nations to make nuclear weapons sooner than they otherwise would. As Mr Morland commented in *Newsweek*: "The legal case against publication is absurd . . . It is based on the proposition that national security would be compromised by revelation of a "secret" that can be expressed in a *single sentence*. I first heard that sentence a year ago, in the University of Alabama, when an anonymous student explained the concept that three Carter administration officers now insist is a vital national secret."

Sure enough, the case soon had to be dropped by an embarrassed administration. Mr Morland was in fact a latecomer to the scene. Two detailed articles, with diagrams — one in a University of Wisconsin underground paper, the other in a New York feminist newspaper — had already explained how to make a mini-nuke with plastic explosives and a caterers'-size coffee-tin.

In Britain, where the nuclear industry is so much less open to public scrutiny, there is as yet little concern about MUF. Flowers noted (paras 323–324): "The construction of a nuclear bomb by a terrorist group would certainly present considerable difficulties and dangers to those attempting it. The equipment required would not be significantly more elaborate than that already used by criminal groups engaged in the illicit manufacture of heroin, but great care would need to be taken in the handling of dangerous materials and in avoiding dangerous criticality. A substantial knowledge would be needed of the physical and chemical processes involved, of the properties of high explosives and of the principles of bomb construction. We have been impressed and disturbed by the extent to which information on all these topics is now available in open technical literature . . . We have concluded therefore that it is entirely credible that plutonium in the requisite amounts could be made into crude but very effective weapons that would be transportable in a small vehicle. The threat to explode such a weapon unless certain conditions were met would constitute nuclear blackmail and would present any government with an appalling dilemma. We are by no means convinced that the British government has realised the full implications of this issue."

Sir Brian Flowers (now Lord Flowers), himself a nuclear scientist and a member of the AEA, personally reinforced this view at the National Energy Conference on 22 June 1976: "I do not believe that it is a question of whether someone will deliberately acquire plutonium for purposes of terror or blackmail, but only of when and how often." And in May 1977, Professor Hans Bethe, arguing *for* nuclear power at the Strasbourg Energy Conference, stated: "Protection against clandestine diversion cannot be achieved by any accounting method now known." (When will someone write a thesis on the psychology of nuclear euphemisms? Why does "theft" have to be "clandestine diversion"?)

The American Electric Power Research Institute and the UKAEA have both admitted that nuclear reactors produce enormous amounts of waste that can be made into bombs; and that nuclear waste storage tanks could become increasingly tempting "plutonium mines" as the radioactivity in spent fuel rods lessens. To deter terrorists, and the agents of non-nuclear states, they suggest a new processing method which would not separate and purify the uranium and plutonium but would combine these elements with other fission products, thus concocting a mix too radioactive for safe handling, should it be "diverted" at any stage. But various techniques of countering this "spiking" precaution are now rapidly being developed. Anyway, are nuke-greedy governments or fanatical terrorists likely to be deterred by the threat of radiation? Many people doubt this. Terrorists are — almost by definition — indifferent to personal safety. True, Louis-François Duchêne has argued that for political reasons they are unlikely ever to "go nuclear". "Urban guerillas basically need to appeal to mass opinion, or a sufficient section of it, and cannot do so by indiscriminate outrage."[14] But events in Ireland alone prove that the general public's "indiscriminate outrage" is the terrorists' triumph.

For many years the accessibility of nukes was suppressed. Professor Albert Wohlstetter, one of the world's leading authorities on nuke-making, who has advised the US government for many years, created a sensation at the Windscale Inquiry by revealing that in 1946 Oppenheimer wrote a memorandum warning that simple atomic bombs could be made from *impure* plutonium. Had this information not at once been sat on, Eisenhower might never have launched his disastrous "Atoms For Peace" programme.

There is no "nuclear weapons secret", and hasn't been for decades; so American governmental support for the Runaway Reactor trade is inexplicable — unless one sees it as an essential part of an all-out effort to rescue America's nuclear power industry from economic collapse. As Ralph Nader has put it:[15] "Nuclear power is in many ways this country's 'technological Vietnam'. The reactor vendors, utilities, engineering construction firms, energy companies, and others have invested many billions of dollars in nuclear technology. So have government officials invested billions of tax dollars in subsidising the development of this energy form over the years. Many scientists, engineers and administrators have invested their careers in highly specialised nuclear operations . . ." And yet — the two leading producers, General Electric and Westinghouse, have had not a single new domestic order since 1975.

The main reasons for this are soaring costs, a dramatic downward revision of forecasts for electricity demand and growing public opposition. A recent analysis[16] has shown that "during 1971–8, real capital cost per installed kilowatt increased more than twice as fast for nuclear as for coal plants and already exceeds the latter by 50 per cent, despite investments that decreased coal plants' air pollution by almost two-thirds and will soon have done so by nine-tenths. For nuclear plants now starting construction — *excluding* the possible impact of tighter federal regulatory standards in the wake of TMI — nuclear capital costs will exceed those of coal by 75 per cent, indicating that many of the ninety US nuclear units with construction permits could be converted to coal to provide cheaper electricity."

A *Foreign Affairs* article[17] further explains: "Unexpectedly high estimated costs for waste management, decommissioning nuclear plants after at most a few decades, and cleaning up past mistakes (for example, burying the hazardous tailings left over from uranium mining) add many billions of dollars in liabilities. Erratic reactor performance — poor reliability, cracks in key components, maintenance problems seeming to go with scarcely a pause from the pediatric to the geriatric — has afflicted most countries. And as cumulative losses mount into the billions of dollars, no vendor in the world appears to have made a nickel on total reactor sales."

Hence "Runaway Reactors", heavily subsidised by the government. These are built with US technology and exported to developing countries where, even if the citizens were sufficiently aware to protest effectively, their leaders would not hesitate to silence them. The terms of a US government-insured loan are so favourable that cost is no great obstacle and the US Export-Import Bank is spending more than $1 billion on supporting these exports. Contracts are easily negotiated, bribery seems to be commonplace[18] and, as environmental impact statements and safety reports are not required, the granting of export licences becomes a mere formality. Thus the US government is guilty of cynical economic imperialism at its most contemptible. Quite apart from proliferation — when hundreds of human-error accidents are known to occur in the US, what is the likely ultimate fate of reactors operated by people with little or no technological tradition behind them?

In 1972, Clifford Beck reviewed foreign reactors for the AEC. He found India's Tarapur reactor — built by two US firms, General Electric and the Bechtel Corporation — "beset with a variety of problems". Because fuel shipped from the US had been leaking radioactivity, the plant's liquid and gaseous effluents were abnormally radioactive and radioactivity had been found in the bodies of local fish-eaters in coastal villages. Within the plant, 1,300 workers had already received their maximum radiation doses. And the radioactive liquid waste system was being operated by Indians *squatting in the rafters of the plant, using long bamboo poles*. As a result of Beck's report, Stephen H. Hanauer of the AEC told a Bechtel executive: "Tarapur is a prime candidate for a major nuclear disaster."[19] No wonder so many American anti-nukes feel illogically guilty about the Runaway Reactor trade; they have helped to check nuclear madness at home, only to see the contagion spreading to countries which are vulnerable to all its worst effects — and have no antidote in the form of a nuke-educated public opinion.

According to David Rosenbaum, a Pandora's Box has been opened and there can be no return to a world free of nuclear weapons. Or, as Lord Carver has expressed it: "You cannot put the nuclear imp back in the bottle." However, it *is* possible to reseal Pandora's Box, or the bottle, to make further nuclear proliferation much more difficult. If the world's nuclear power industrialists admitted their mistake *now* — as they will certainly have to do one day, if only for "market-place" reasons — what would be the inhibiting consequences for nuke-makers? Again, *Foreign Affairs* explains:[20] "With trivial exceptions, *every* known civilian route to bombs involves *either* nuclear power *or* materials and technologies whose possession, indeed whose existence in commerce, is a direct and essential consequence of nuclear fission power . . . Suppose that nuclear power no longer existed. There would no longer be any innocent justification for uranium mining (its minor non-nuclear uses are all substitutable), nor for possession of ancillary equipment such as research reactors and critical assemblies, nor for commerce in nuclear-grade graphite and beryllium, hafnium-free zirconium, tritium, lithium-6, more than gram quantities of deuterium, most nuclear instrumentation — the whole panoply of goods and services that provides such diverse routes to bombs. If these exotic items were no longer commercially available, they would be much harder to obtain; efforts to obtain them would be much more conspicuous; and such efforts, if detected, would carry a high political cost because for the first time they would be *unambiguously military* in intent . . . This does not absolutely mean that a determined and resourceful nation bent on bombs can by non-military means be absolutely prevented from getting them: much is already out of the barn. But denuclearisation would narrow the proliferative field to exclude the vast majority of states — the latent proliferators who sidle up to the nuclear threshold by degrees."

7

MAD

The NATO doctrine is that we will fight with conventional forces until we are losing, then we will fight with tactical nuclear weapons until we are losing, and then we will blow up the world.
MORTON HALPERIN, formerly US DEPUTY ASSISTANT-SECRETARY OF DEFENSE[1]

One of the most staggering and scandalous facts of the nuclear age has been the arrogance with which the leading statesmen of the world have tried to keep within tiny cliques the truth about a matter of such supreme consequence to everybody.
MICHAEL FOOT, 1964[2]

The menace of nuclear war reaches far back into the economies of both blocs, dictating priorities and awarding power . . . Here also will be found the driving rationale for expansionist programmes in unsafe nuclear energy, programmes which cohabit comfortably with military nuclear technology whereas the urgent research into safe energy supplies from sun, wind or wave are neglected because they have no military pay-off.
E. P. THOMPSON, 1980[3]

My original intention was to avoid the weapons debate, lest it might confuse my argument against the spread of nuclear power. But one cannot for long escape the fact that nuclear weapons and nuclear power are two sides of the same coin and are being forced on us by the same intercontinental industrial/military/political machine (IMP — to coin another personal acronym). This machine runs on different fuels in the Capitalist and Communist worlds but everywhere its function is the same: it is an anti-human bulldozer. Yet many of its operators are delightful individuals who know not what they do — which makes it that much more difficult to stop the machine.

Looking around the world now, it is hard to believe that thirty-five short years ago the nuclear arms race was *not* inevitable. Eleven days after Hiroshima, the Science Panel of the US Interim Committee on the use of nuclear energy — headed by J. Robert Oppenheimer, "Father of the Atomic Bomb" — sent to Henry L. Stimson, Secretary for War, some advice about contemplated legislation on atomic development. The scientists pointed out that there could never be any satisfactory defence against nuclear weapons. They expressed grave doubts about their defensive value for any nation and argued that American attempts to keep the nuclear secret, instead of sharing it with Russia, could only encourage mistrust and competition. They urged their government to initiate the setting-up of an international body to ban the production of nuclear weapons and control the use of nuclear energy. Stimson agreed with all this but James Byrne, Secretary of State, told Oppenheimer that to make any such international arrangements would simply not be practical. Oppenheimer, as Director of Los Alamos Laboratory, was ordered to continue r & d on America's nuclear arsenal.

On 12 September Stimson presented a memorandum of his own views to Truman:

In many quarters the atomic bomb has been interpreted as a substantial offset to the growth of Russian influence on the Continent. We can be certain that the Soviet government has sensed this tendency and the temptation will be strong for the Soviet political and military leaders to acquire this weapon in the shortest possible time . . . Accordingly unless the Soviets are voluntarily invited into a partnership upon a basis of cooperation and trust, we are going to maintain the Anglo-Saxon bloc over against the Soviet in the possession of this weapon. Such a condition will almost certainly stimulate feverish activity on the part of the Soviets towards the development of this bomb in what will in effect be a secret armaments race of a rather desperate character. There is evidence to indicate that such activity may have already commenced . . . Our relations may be perhaps irretrievably embittered by the way in which we approach this solution of the bomb with Russia. For if we fail to approach them now, and merely continue to negotiate with them, having this weapon rather ostentatiously on our hip, their suspicions and their distrust of our motives and purposes will increase. If the atomic bomb were merely another though more devastating military weapon to be assimilated into our pattern of international relations it would be one thing. We could then follow the old custom of secrecy and nationalistic military superiority, relying on international caution to proscribe the further use of the weapon as we did with gas. But I think the bomb instead constitutes merely a first step in a new control by man of the forces of nature, too revolutionary and dangerous to fit into the old concepts. I think it really caps the climax of the race between man's growing technological powers of destruction and his psychological power of self-control and group control — his moral power. If so, our method of approach to the Russians is a question of the most vital importance in the evolution of human progress.

Truman read all this aloud in Stimson's presence and appeared to agree with it. But later he wrote in his memoirs that long before this: "I had decided that the secret of the manufacture of the weapon would remain a secret with us." The Soviet army had recently moved into Jachimov and St Joachimstal and occupied the uranium factory there — the only place in Europe where uranium was produced — in defiance of an agreement with Czechoslovakia. And Russia was also being predatory in the Far East: so Truman saw exclusive possession of the bomb as too valuable a political lever to waste.

This tragic decision is understandable. Men of vision, like Stimson, are necessarily always in the minority and Truman had to do the best he could wearing those blinkers which are part of the average politician's equipment. Yet surely there is a lesson here for the politicians of our own generation. Because Truman and his advisers chose the obvious, immediate advantage, mankind is now threatened with extinction. Nuclear power, as Stimson pointed out, is too dangerous to fit into the old contexts. But that is precisely what IMP is still trying to do, thirty-five years on: fit new weapons into old contexts. Can they not learn from Truman's mistake before it is too late? Or is the IMP bulldozer now so out of control that its own operators are unable to stop it?

On 29 October 1949 the General Advisory Committee of the AEC (which had not yet been taken over by the hawks) voted unanimously against developing the H-bomb: "We all hope that by one means or another the development of these weapons can be avoided. We are all reluctant to see the US take the initiative in this development. We are all agreed that it would be wrong at the moment to commit ourselves to an all-out effort towards its development . . . In determining not to proceed to develop the

Super bomb, we see a unique opportunity of providing by example some limitations to the totality of war and thus eliminating the fear and arousing the hope of mankind."

Enrico Fermi and I. I. Rabi added a minority report: "The fact that no limits exist to the destructiveness of this weapon makes its very existence and the knowledge of its construction a danger to humanity as a whole. It is necessarily an evil thing, considered in any light. For these reasons, we believe it important for the President of the United States to tell the American people and the world that we think it wrong on fundamental ethical principles to initiate the development of such a weapon."

Strong words — yet IMP prevailed. It had to have "weapon superiority" and recked nothing of Good and Evil. The decision to go ahead was precipitated in January 1950 when Klaus Fuchs, the German scientist who had worked on the Manhattan Project, was exposed as a Communist spy. According to Teller, the "Father of the H-bomb" who longed for an excuse to develop it, Fuchs had been able to give Russia almost complete knowledge of American nuclear activity since June 1944. Four days after Fuchs's arrest, Truman rejected the scientists' final plea for restraint and went ahead with the thermonuclear programme.

On 1 November 1952 America detonated a fusion bomb in the South Pacific: nine months later Russia did the same. By November 1958, the Americans and the Russians in their respective tests had detonated more than one hundred times the explosive power dropped on Germany during World War Two. In 1960 one American university accepted $40,000,000 in defence research contracts, with forty other universities getting a million dollars each. War had become Big Biz and it has been getting bigger ever since. By now military research is absorbing the creative powers of more than 600,000 scientists and engineers throughout the world and it is receiving more public funds than all social needs combined. The military/industrial complex employs more than half of America's scientists. And, for all the talk about an "energy crisis", weapons research gets six times as much funding as energy research — including nuclear power, which naturally is allocated by far the greater part of the energy research budget.[4]

Ruth Leger Sivard, a former chief of the economics division of the US Arms Control and Disarmament Agency, has written: "There is in this balance of global priorities an alarming air of unreality. It suggests two worlds operating independently of each other. The military world, which seems to dominate the power structure, has first call on money and other resources, creates and gets the most advanced technology, and is seemingly out of touch with those threats to the social order which have nothing to do with weapons. The other world, the reality around us ... is a global community whose members are increasingly dependent on one another for scarce resources, clean air and water, mutual survival. Its basic problems are too real, too complex, for military solutions."[5]

IMP is not easily dissected. Does the military world dominate the power structure, as Dr Sivard suggests? Or are industrialists and scientists now in control of the military? Has the Pentagon got so hooked on technological novelties that no military expert can say to the inventor of the latest homicidal device, "No, we don't *need* that — take it away!"? We can only know for sure that IMP is the most menacing alliance ever to have evolved within human society.

During the past few years IMP's inhumanity has been repeatedly exposed in America, where the Freedom of Information Act can now be used to unlock thousands of skeleton-packed cupboards. The AEC records, in particular, reveal a remarkably ruthless organisation. The Commissioners were determined to "perfect" America's nuclear weapons as soon as possible, regardless of public health consequences, and in 1951 they began their long campaign of suppressing inconvenient research results. In March of that year Arnold B. Grobman published *Our Atomic Heritage*, in which he criticised many AEC shortcomings, including their failure adequately to protect laboratory workers from radiation. Even before the book appeared the AEC had planned a clever and well-financed attack on his reputation and thus he became the first of their many victims. During this period they also perfected the art of misleading the public, which skill has since, alas! been acquired by the UKAEA.[6]

The Pentagon shows up no better. Between 1948 and 1963 almost 300,000 US troops were exposed to low-level radiation — inefficiently monitored — during hundreds of bomb tests conducted jointly by the Pentagon and the AEC. Amongst these soldiers, the incidence of various types of leukemia is now more than double the national average. Yet they are unable to obtain compensation. The Veterans' Administration stubbornly denies any possible link between their leukemia and their exposure to low-level radiation; to admit that there could be such a link would endanger the nuclear power industry, which has taken so many hard knocks during the '70s, quite apart from TMI.[7]

The iodine-131 tragedy showed the AEC at their most brutal. In 1963, Dr Eric Reiss, Professor of Medicine at St Louis's Washington University, reported at a public hearing that the amount of iodine-131 in milk produced near the Nevada test site was at levels considered carcinogenic even by the notoriously slap-happy Federal Radiation Council (FRC). Yet no attempt was made to halt consumption of the contaminated milk. Instead, a few months later, the FRC decided secretly on a *threefold* increase in the allowable dose of radioiodine in milk. This meant that health inspectors did not have to strengthen public suspicions about low-level radiation by taking contaminated milk off the market.[8]

Two years later Dr Weiss of the Bureau of Radiological Health announced his findings on the abnormally high rate of thyroid cancer — a direct consequence of drinking contaminated milk — among Utah children. Much of his data was at once destroyed by the AEC. Internal AEC documents dating from 1965 and published in 1979 revealed that the Commissioners of the time suppressed Dr Weiss's report because they wished "to avoid jeopardising the development of the nuclear power industry".[9]

IMP consistently chooses "Profit Before People", both at home and in the Third World. Yet on these islands we remain deeply reluctant to face up to the possibility of such extreme governmental immorality infecting our own countries. This is why I have given several examples of the symptoms we may expect to notice, as traditional standards are eroded by IMP's nuclear megalomania. (There are of course many more examples on the records.) Even in Britain there are disquieting signs that IMP methods are percolating through, behind the nuke scenes. The secret "Chevaline" programme to modernise Britain's Polaris missile warheads cost £1,000 million and was decided upon while the Labour Party was in power, *contrary* to official Labour policy, which was opposed to any replacement of the whole Polaris system, *without* the knowledge of the full Cabinet (never mind Parliament) and on the authority *only* of Messrs Callaghan, Healey, Owen and Mulley. E. P. Thompson has commented: "I do not know how a thousand million was tucked away in a crease in the estimates and hidden from view, but it suggests that the level of official mendacity is today very high indeed."[10]

Britain's nuclear policies are now being decided at secret NATO committee meetings. On 12 December 1979 one such meeting decreed that 160 or so US cruise missiles should be stationed in Britain during the '80s — all of course to be "owned and operated" by the US. American military leaders have openly spoken of firing those missiles from Britain, entirely at their own discretion. But perhaps it's best that such a decision should be taken unilaterally. If consultations were required, things could get awkward at the crucial moment. One is not all the time aware of Mrs Thatcher's tenderheartedness, yet she *is* a woman and even she might hesitate if asked to participate in button-pushing that could cause some twenty-five million of her compatriots to die horribly if retaliation came. And should she hesitate, what then? Any American soldier who, overcome by silly sentimentality, refused to button-push on order, must at once be shot by a comrade; but presumably an irresolute Mrs Thatcher, if one can conceive of such a metamorphosis, would be dealt with less drastically. Still, precious minutes would have been lost, during which those svelte missiles could have covered hundreds of miles. So doubtless it is more practical, from the military point of view, to leave the fate of Britain in American hands.

Field-Marshal Lord Carver gave his personal view of the new cruise missile in the course of a lecture on "Nuclear Weapons in Europe", delivered to the Welsh Centre for International Affairs at Cardiff on 17 October 1980.

In many ways I see it as a more satisfactory nuclear weapon delivery system than aircraft. Apart from being cheaper and avoiding the need to send aircrew on a very hazardous mission, it does not need, and would not be fired from, a limited number of fixed bases, like aircraft. Aircraft bases would be high priority targets for enemy nuclear attack, and pre-emptive attacks to knock them out would be a temptation to the enemy as one of the first blows to be struck in any war. I believe, therefore, that the proposal to base cruise-missiles in this country and elsewhere in NATO should not be opposed — indeed, if they were to replace aircraft bases, they should be positively welcomed. I find it difficult to give them an unqualified welcome on other grounds, for two reasons. First, that they represent yet another development in the nuclear weapons systems arms race; and secondly, because I have reservations about the general concept that it is necessary or desirable to establish an exact balance, or even a superiority in numbers and quality, between the European theatre nuclear weapons amories of NATO and the Warsaw Pact . . . The concept of such a balance encourages the idea that a nuclear war could be carried on between NATO and the Warsaw Pact, limited to theatre nuclear weapons within Europe, the homelands of the United States and the Soviet Union remaining unharmed. That may be a nice idea for them, but not for Europe.

On unilateral nuclear disarmament, Lord Carver said:

To give up our own nuclear weapons would have a political effect in Europe and in the world as a whole which would be very serious and disadvantageous to the West. It would look as if we were opting out of defending the interests of the West. It would be a very positive decision with immense implications. So I am in favour of continuing to have our own nuclear weapons, but not in favour of continuing to maintain what we call our independent strategic deterrent . . . The general concept that I favour, regarding nuclear weapons in Europe, is to make it as clear as possible to anyone who might be tempted to embark on war in Europe, first of all, that they will have to overcome the stubborn and efficient resistance of the conventional forces of NATO, kept ready and able to confront them. Secondly, that, if they persist in their attempts to overcome NATO's conventional forces, and certainly if they use nuclear weapons in their attempt to do so, they will incur the risk of nuclear attack on their forces, which are being used and deployed for that purpose. Thirdly, if that does not bring a halt to their aggression, that they incur the risk of direct attack on their homeland and the unleashing of a major nuclear war, including attacks on cities, which has rightly been described as a holocaust. This is a high risk policy. But as long as the risk of going to war is high, the temptation to embark on it is low. A low risk policy in the short term produces high risks in the longer term. Appeasement in the 1930s was such a low risk policy, and it led to World War Two. Pursuing a low risk policy now would probably lead to World War Three . . . We should aim at a stable military balance, reflecting a stable political balance

. . . It must include nuclear weapons; but it need not, and it should not include them in the ridiculously and dangerously large numbers that it now does. It could be at a very much lower level. All our efforts should be bent, first, towards establishing a stable balance at the existing level: then to reducing it, while keeping it stable. If, in that process, we can contribute to its success by giving up our own nuclear weapons, I would favour that; but, at present, it would not help progress, perhaps indeed hinder it. Progress will demand infinite patience, clear vision and a degree of dispassionate acceptance of unpleasant realities that it is rare to find in any body of men and women, bodies politic or other.

Lord Carver is nobody's idea of a hawk, and the fact that he writes good plain English accentuates the horror of what he has to say. Morton Halperin's assertion at the head of this chapter seems like the raving of some paranoid Communist-basher — yet Lord Carver reinforces it, using more restrained language: "If that does not bring a halt to their aggression . . . they incur the risk of unleashing a major nuclear war . . ." Or, as Halperin put it, "Then we will blow up the world." From what Lord Carver has said, we must deduce that the possibility of a holocaust is one of the unpleasant realities to be dispassionately accepted while slow progress is being made towards the reduction of nuclear weapons in the superpowers' stockpiles — and indeed long after that, since *reduction* is not *elimination*.

The nuclear deterrent is no deterrent unless those in charge make it plain that they are prepared to use it if attacked. And no one can dispute its effectiveness, to date. In a nukeless world, a war between the superpowers, killing millions, might well have flared up during any of the several major crises of the past thirty years. But can such a mutual-terrorisation method of maintaining peace continue to work indefinitely? Is it consistent with military human nature to possess, but *never* use, increasing numbers of brilliantly sophisticated weapons? During the '70s, the designers of nuclear weapons systems reached ever higher levels of ingenuity and subtlety, thus tempting some military planners to stray from the path of *pure* deterrence. Deterrence has worked so far partly because the results of using old-fashioned maxi-nukes are "unacceptable" by anyone's definition of that word. But the results of using the new-fangled mini-nukes seem, to some military experts, "acceptable". And so these men have persuaded themselves, though not many others, that it is feasible to use mini-nukes only, without unleashing maxis. Hence "the erosion of deterrence" is being widely and nervously discussed, as we are told that plans for a "limited" nuclear war must be part of the "Western flexible response".

But what can justify using nuclear weapons of any type or size? NATO answers: "The avoidance of the defeat of our conventional forces by Russia." So to *protect* our area of the world, they claim to be willing, if cornered, to use weapons which would almost inevitably result in the destruction of *most* of the world — *including* our own area. The fundamental irrationality here is quite petrifying; if people are sufficiently off-beam to argue thus they may well be incapable, under certain conditions, of controlling their war-machine. Many consider NATO's attitude immoral but I disagree: to be immoral one must be in full possession of one's senses, which our "protectors" demonstrably are not. True, they insist that they hope never *ever* to have to use even the *teeniest* nuke and one doesn't doubt this. Nevertheless, they also state that in certain circumstances they would *initiate* a nuclear exchange. They no longer regard nukes exclusively as deterrents, to be used only in retaliation for a nuking of their own territory. Instead, they are thinking and talking of them as though they were on a par with non-nuclear weapons and could reasonably be used to back up conventional methods of attack and defence against conventional forces ... No wonder nuclear war *feels* nearer now than it did a few years ago; what for decades was unthinkable has suddenly become "thinkable". So how reassuring is the argument that: "As long as the risk of going to war is high, the temptation to embark on it is low?"

Common-sense revolts against the notion that we can go on from decade to decade, spending more and more on weapons of *deterrence* when already the superpowers possess a gross superfluity of nukes — enough to deter a whole galaxy. Lord Carver has said: "A stable military balance must include nuclear weapons; but it need not, and it should not, include them in the ridiculously large numbers that it now does." So — are IMP's continuing r & d programmes really being run for the benefit of those experts whose skills could find no outlet in a peaceful world? Can it be that everybody is bluffing all the time and nobody anywhere has the slightest intention of ever pushing a button? Is "nuclear deterrence" just a sick, cruel game which to date has cost the superpowers some $500 billion in public funds, and is still costing over $40 billion a year? In the US, "Defense" now absorbs almost one-third of all Federal revenues. And President Reagan's increased military expenditure means that for each man, woman and child in America $1,200 will be spent annually on arms until 1985.

The nuclear arms race may indeed have been a game of bluff during the '60s, but it is no longer so. Nor, at present, is IMP-ish r & d merely a Welfare Scheme for scientists who might otherwise be on the dole — as it perhaps once was. IMP is now striving to perfect mini-nukes. And, once these have been perfected, how long will it take for nuclear numbing fatally to distort the perceptions of some military planner or political leader?

Apart from the direct threat inherent in the doctrine of "flexible response", we cannot hope to get away indefinitely with isolating ourselves from the rest of the world, and concentrating on "keeping a stable military balance" in Europe, while the global sum of human misery daily increases. Sooner or later the Great Unfed will be at our throats, with every excuse, and we will be out of the Soviet frying-pan into the Third World fire. And anyway, what sort of peace is this of which our military mentors boast? It is based on threat and counter-threat, on terror and suspicion and propaganda-fostered hates. To speak of "Europe's thirty-five years of peace" is an almost blasphemous abuse of the word. True peace is not merely an absence of aggressive military action. It is a state of mind — a set of attitudes — that would utterly destroy the IMP monolith if it were allowed to spread.

Lord Carver thinks that if Britain disarmed unilaterally "it would look as if we were opting out of defending the interests of the West . . . It would be a very positive decision with immense implications." I see other implications. There cannot be one human being, in any country, who *wants* a nuclear war; yet no world leader dares make a move towards sanity and disarmament lest "the balance" be upset. But if Britain disarmed this would not, in fact, upset the balance. Both sides are so prodigiously over-armed that her nukes, plus whatever the Americans may have stationed in or around Britain, are of no real significance. Their absence would mean that the war-game had to be played slightly differently but the superpowers could still destroy each other (and the rest of us) dozens of times over. Meanwhile, Britain nuke-free could inspire a world that longs for leadership based not on expediency and fear but on courage and hope. Her example might do more than we can now imagine to deliver us from evil; there are intangible forces in this world, as well as military ones.

Joseph Rotblat, until recently Professor of Physics in the University of London, worked as a young man on the Manhattan Project but dropped out when it became clear that Germany was not making the bomb. In 1976 he wrote: "The attitude towards nuclear war is becoming similar to that towards potential natural disasters: we know the threats exist, but there is nothing we can do about them. This fatalistic attitude is, of course, fallacious and unrealistic. Unlike Acts of God, which by definition are unpredictable, the occurrence of a nuclear war is a predictable event; its probability is increasing with time. The so-called nuclear stalemate is not a static phenomenon; it is dynamic, and can only be described as a state of unstable equilibrium with a pre-determined outcome . . . Our only hope of survival is a return to the earlier concept of bringing the arms race to an end by general and complete disarmament."[11]

Meanwhile IMP's minions are jollying us along to go with the new tide in the affairs of strategists. On 12 August 1980, Mr Douglas Hurd, Minister of State at the Foreign Office, blandly announced on television that "the view shared by certain military experts that nuclear war could be conducted without vast loss in civilian lives was not dangerous and was probably realistic".

Lord Mountbatten disagreed. A few months before his murder he condemned the arms race in general, and the notion of "theatre" warfare in particular, at a meeting at Strasbourg on 11 May 1979, to mark the awarding of the Louise Weiss Foundation Prize to the Stockholm International Peace Research Institute. In a speech which was given strangely little space in the British press he said: "The belief now is that small nuclear weapons could be used in field warfare without triggering off an all-out nuclear exchange leading to the final holocaust. I have never found this idea credible. I know how impossible it is to pursue military operations in accordance with fixed plans and agreements. In warfare the unexpected is the rule, and no one can anticipate what an opponent's reaction will be to the unexpected. I can see no use for nuclear weapons which would not end in escalation, with consequences that no one can conceive. And nuclear devastation is not science fiction — it is a matter of fact. As a military man who has given half a century of active service, I say in all sincerity that the nuclear arms race has no military purpose. Wars cannot be fought with nuclear weapons. Their existence only adds to our perils because of the illusions which they have generated. There are powerful voices around the world who still give credence to the old Roman precept: 'If you desire peace, prepare for war.' This is absolute nuclear nonsense: it is a disastrous misconception to believe that by increasing the total uncertainty one increases one's own certainty."[12]

B. H. Liddell Hart has written: "It is quite untrue that if one wishes for peace one should prepare for war, but if one wishes for peace one should understand war."[13] This advice prompted me to tackle a collection of essays entitled *Strategic Thought in the Nuclear Age* and edited by Laurence Martin (Heinemann, 1980).

In an essay on "Disarmament and Arms Control Since 1945", John Garnett cheerfully admits that the annual expenditure on armaments is "about equal to the entire national income of the poorer half of mankind, and the amount given in aid to underdeveloped countries is a mere five per cent of the money spent for military purposes". Nevertheless, Mr Garnett considers it folly to support disarmament: "Those, like Bertrand Russell or Alva Myrdal, who fulminate about the 'reign of unreason' or 'global folly', make the mistake of judging the arms race from the detached 'god-like' perspective of a rational, omniscient being looking down on planet earth. To such a being, the arms race, like many other aspects of human behaviour, must look absurd. But criticism from that perspective is as irrelevant to an individual statesman as human criticism of the behaviour of lemmings is to an individual lemming." This curious analogy scarcely flatters individual statesmen: nor does it inspire much confidence in Mr Garnett's perspicacity. Yet he is a most distinguished man in his own field: among other things, a member of the Foreign and Commonwealth Office Advisory Panel on Arms Control and Disarmament and a British member of the UN Advisory Board on Disarmament Studies. His views have to be taken seriously; he is one of those decision-makers who mould the world we live in.

John Garnett concedes *en passant* that "large numbers of people have a vested interest in perpetuating the arms race". He also points out, reasonably enough, that "the causes of international tension and war lie much deeper than the machinations of either individuals or even powerful pressure groups". But he ignores the extent to which international tensions are now being heightened by governments with an interest in promoting arms sales. War may well be the oldest profession — though another lays claim to that distinction — but the arms trade is a new and peculiarly vicious feature of our existence. The Lockheed scandal affords a classic example of how multinationals operate, with the furtive support of governments, when billions of dollars are at stake.

IMP was born over a century ago, when Science and Technology were recruited to modernise Western armies. The relevant industries, untrammelled by even the flimsiest moral restraint, rapidly gained strength. In Britain, Sir Basil Zakaroff, chief salesman of Vickers, sold arms to both sides during the Boer War and World War One. Krupps in Germany were no better and, as such companies acquired more and more industrial and financial power, so their political influence increased. Since the end of World War Two the superpower governments have lavished armaments on Third World countries, in attempts to gain or maintain regional influence, and the industries concerned have prospered — while over a hundred wars have been fought in some sixty countries. Ex-colonial powers eagerly exploit the longings of newly independent nations for the most expensive tanks and war-planes; France and Britain, for instance, welcome young Asian and African officers to their military academies as a way of promoting future arms sales. And both America and Russia have used the latest weapons as bribes to secure military bases.[14]

In 1921 a League of Nations report complained: "Wars are promoted by the competitive zeal of private armaments firms." And in our day — according to Michael Klare, a foreign policy analyst at Princeton University's Centre of International Studies — Pentagon officials are told that their job of increasing arms sales should be regarded as "a mission".[15] So they stalk the earth in search of prospective customers, accompanied by the corporate executives of arms manufacturers. (And occasionally preceded, in the few tranquil areas of the world, by anonymous characters with a curious facility for arousing cross-border suspicions.)

Between 1968 and 1975, US arms exports rose by over 1,200 per cent and the pace continues to quicken. America, the world's main arms supplier, provided 53 per cent of Third World arms during those years. Since 1975, orders for new arms have exceeded deliveries by more than three to one — a spiral that must have gratified a recent Deputy-Secretary of Defense, William Clements, who has said: "Any restriction on weapons exports decreases the potential contribution of sales to strengthening both free world security and the US economy and balance of payments position." IMP is rarely so frank.[16]

Throughout the '70s, Britain's arms sales became increasingly important. In November 1976 Prince Sultan of Saudi Arabia arrived in London to be greeted by a guard of honour and an ecstatic Minister of Defence — Fred Mulley — while in the background Rolls-Royce and the British Aircraft Corporation bit their nails and wondered about the size of the order. Presumably no one mentioned John Stonehouse, who eleven years previously had visited Saudi Arabia to negotiate Britain's biggest-ever export deal with the arms-loving Prince Sultan.

Ruth Leger Sivard has pointed out that such deals have snags: "The trade in arms is promoted with government assistance. Although considered an offset to escalating costs of oil and other raw materials, the major portion of this trade goes to developing countries, where it inevitably puts pressure on the prices they charge for oil and their raw materials — the spiral is unending."[17] And from the purchaser's point of view the long-term disadvantages are still more serious: "Developing countries are affected by the diversion of labour and management skills to military programmes, and to the advanced technology that increasingly goes with them. They have fewer trained people to spare. Military requirements drain away talent essential for development. They may also introduce at too early a stage the complex technology that can be paralyzing for young economies."[18]

However, Britain at least keeps the party clean. Carl Kotchian, one of Lockheed's more celebrated salesmen, once remarked that he could always smell corruption and he smelt it in most countries — apart from Britain, Canada, New Zealand and a few others.[19]

In May 1976, speaking in Parliament, Roy Mason complained angrily: "If the MRCA contract is cancelled . . . 24,000 jobs would go immediately . . . There would be hardly any aircraft industry left." Yet a year earlier the Lucas Aerospace workers had themselves suggested many alternative peaceful uses for their factories, including the production of kidney machines, devices to help spina-bifida children, special electronic aids for operating theatres and large-scale windmills for electricity generation. Their Corporate Plan comments: "There is now deep-rooted cynicism amongst wide sections of the public about the idea, carefully nurtured by the media, that advanced science and technology will solve all our material problems. Of particular significance in this connection is the much publicised rejection by capable sixth-formers of the places that are available for science and technology at British universities. Science and technology are perceived by them to be de-humanised and even brutal and the source of a whole range of problems, not only for those who work in the industries themselves but also for society at large. It is our view that these problems arise, not because of the behaviour of scientists and technologists in isolation, but because of the manner in which society misuses this skill and ability. We believe, however, that scientists, engineers and the workers in those industries have a profound responsibility to challenge the underlying assumptions of large-scale industry and to assert their right to use their skill in the interests of the community at large. In saying that, we recognise that this is a fundamental challenge to many of the economic and ideological assumptions of our society . . . Our intentions are to make a humble start to question these assumptions and to make a small contribution to demonstrating that workers are prepared to press for the right to work on products which actually help to solve human problems rather than to create them."[20]

John Garnett would have no patience with those ideals. He quotes Salvador de Madariaga: "The problem of disarmament is not the problem of disarmament. It really is the problem of the organisation of the world community." In Mr Garnett's view: "The system change implied by de Madariaga's accurate diagnosis is beyond us." He agrees with Hedley Bull, who has "rightly condemned such impractical solutions as 'a corruption of thinking about international relations and a distraction from its proper concerns' ". Happily, however, the Lucas Aerospace workers are not alone in realising that the true "corruption of thinking" has occurred within IMP. Even the Russians are concerned. In 1976, Stanislav Menshikov, a member of the Soviet UN delegation, admitted in an interview on Radio 3: "Unless our military/industrial beast is tamed, the reduction of armaments and military spending will remain a Utopia."

A long time ago Eisenhower said: "I think people want peace so much that one of these days governments had better get out of their way and let them have it." However, since then IMP has been diligently cultivating distrust and hostility and Russians and Americans have been conditioned to hate each other. In America, IMP-ish propaganda has been so successful that at present even a small unilateral nuclear arms reduction would encounter powerful public opposition. According to a 1978 poll, 48 per cent felt that America must have military superiority over Russia and 42 per cent voted for equality. To achieve this, most people were willing to see an extra $10 billion being spent annually on defence — and to provide it, through higher taxation.

Commenting on that poll, *Nucleus* (the journal of the UCS) wrote: "This indicates a serious public misunderstanding of military logic in the nuclear age: when each country can already destroy the other, even after being attacked first, it is impossible to produce any kind of military superiority (or global security) by building more nuclear weapons."

John Garnett refers to the "inevitability of unceasing conflict in international relations". He goes on: "Ambitious and aggressive men, perhaps driven into competing for a more than fair share of the fruits of the earth by a will to power, find themselves thwarted by the similar ambitions of others and are therefore forced to resort to physical violence as the ultimate arbiter of who is master." But, having acknowledged that "to some extent at least, arms races develop because statesmen see an advantage in using military power to achieve their objectives", Mr Garnett gets in a tangle and goes on in his very next paragraph: "The arms race is not an insane out-of-control suicide race promoted by men who are either stupid or wicked. It is a result of reasoned decisions by sensible men grappling to the best of their ability with the wretched situation in which they find themselves." But who made that situation? And who profits from it? And who steadfastly opposes any attempts to alleviate it?

Bertrand Russell and Alva Myrdal are accused of judging the arms race from a detached "god-like" perspective — but consider the perspective of Robert E. Osgood in his essay on "The Post-War Strategy of Limited War: Before, During and After Vietnam". "In order to strengthen deterrence, and avoid a catastrophic war if deterrence should fail, it was necessary to enhance the nuclear capacity of the US to use tactical and even strategic nuclear weapons within tolerable limits of physical and human destruction for limited ends."

When do nuclear weapons become tolerable from the victims' point of view? IMP's answer is hinted at in Henry S. Rowen's essay on "The Evolution of Nuclear Strategic Doctrine": "Another contingency mentioned by James Schlesinger was the possibility of a Soviet attack limited to US ICBMs and SAC bomber bases: if the Soviets chose to limit collateral damage, the resulting US fatalities could be held to a small fraction of the fatalities produced from direct attack on US cities . . . wherein American fatalities would be around 95–100 million people from prompt effects plus fallout . . . An attack on ICBM silos, SAC bases and ballistic missile submarine bases was estimated as producing 5–6 million fatalities; one on ICBMs alone, about one million; and on SAC bases alone, about 500,000 fatalities. In short, discriminate attacks were becoming feasible." For how much longer must we allow our destinies to be shaped by men who consider tolerable half a million fatalities because it is only "a small fraction" of 95–100 million? It is still half a million dead people.

As my tortured readers will long since have noticed, the inelegance of IMP English matches the feebleness of IMP wits. E. P. Thompson is good on this: "The deformation of culture begins within language itself. It makes possible a disjunction between the rationality and moral sensibility of individual men and women and the effective political and military processes. A certain kind of 'realist' and 'technical' vocabulary effects a closure which seals out the imagination, and prevents the reason from following the most manifest sequence of cause and consequence. It habituates the mind to nuclear holocaust by reducing everything to a flat level of normality. By habituating us to certain expectations, it not only encourages resignation — it also beckons on the event."[21]

These essayists make a number of nervous and irritable references to the power of public opinion in the West. Louis-François Duchêne, for five years Director of the International Institute for Strategic Research, points out: "In Western societies, the process of individualisation has reached a stage where it is hard . . . for the state to mobilise its citizens . . . because of the standards of individual fulfilment growing in the logic of the culture. In the West, the idea of progress has reached a post-patriotic stage when the state is expected to serve the individual, not the individual the state as in the early populist societies. The reluctance of Western Europe to undertake military policies since the war and the reaction, though relatively belated, of American society to intervention in poor countries overseas, are both rooted in this basic development of advanced industrial societies."

So perhaps the sullen, greedy, post-patriotic unrest of our strike-crippled countries is essentially a creative force, helping to undermine IMP by showing that "the growth society" cannot provide a permanently stable basis for civilised living. Those of us old enough to have been brought up "patriotic" cannot help viewing the Scrounger State with some unease; but perhaps it is an unavoidable stage on the way to that New Society which (nukes *volente*) could evolve during the twenty-first century. And it may be that we can escape from the nuke trap only through an anti-IMP revolt (preferably non-violent) of post-patriotic citizens who see that the annual expenditure on armaments vastly diminishes their opportunities for "individual fulfilment" — quite apart from constantly threatening their lives. If Britain were not now spending so prodigiously on defence, would she have to cut back on health services, social security and education? According to Treasury calculations, based on the probable rate of inflation, the cost of her new fleet of nuclear submarines may have risen to £13,000 million by 1990. Do the majority in Britain *want* all these superfluous nukes floating around the place? If they don't, what price democracy?

In 1880 the main function of a European army was to conquer new territory and open up profitable markets. In 1980, as Laurence Martin notes in his essay on "The Role of Military Force in the Nuclear Age", "Conquest has come to appear wrong as well as unprofitable." A century ago, the few who protested against imperial expansion and exploitation seemed freakish and tiresome, just as anti-nukes now do to IMP, when they complain about armies being "the object of more anxious attention and more lavish expenditure than ever before in peacetime". Professor Martin, Vice-Chancellor of the University of Newcastle-upon-Tyne, thoroughly approves of this attention and expenditure and is most critical of "the perverse and pervasive tendency within the democracies that, against overwhelming evidence, persists in treating military activity as an aberration. Widespread public distaste for military affairs, compounded no doubt by increasing personal disinclination from (*sic*) military service in affluent societies, has itself become a major incentive for governments to prefer oblique and indirect strategies of conflict management. Such subtleties have, in turn, reinforced the illusion that military power can safely be neglected."

Professor Martin startles me. If governments are evolving "oblique and indirect strategies of conflict management", instead of going to war, should we not rejoice? Apparently not . . . But *why* not?

Having outlined the doctrine of Mutual Assured Destruction — fittingly known as MAD — Professor Martin concludes: "There is every reason to believe that this is no merely transient phase of history. Despite the once-prevalent endorsements of General and Complete Disarmament as an ultimate goal, despite President Carter's more recent assertion that the abolition of nuclear weapons was his final purpose, the actual historical tendency is to build them ever more firmly into the international system . . . However frequently misunderstood and misrepresented in public, Strategic Arms Limitation is not an alternative to giving nuclear weapons a major role in international politics, but merely one way of defining that role."

In the few years since those words were written, dependence on MAD has in fact been proved a "transient phase of history". Now we have instead the incomparably more dangerous doctrine of "flexible response", which has stimulated a resurgence of nuclear disarmament campaigns throughout the West. The instinct of self-preservation is at work. And the grotesque illogic of IMP's arguments is becoming increasingly apparent to those who have the courage to *stop and think* about this unprecedented crisis in human history.

Military waffle often obscures the fact that if one fights a nuclear war to defend "the interests of the West" there won't *be* any West. On the other hand, should the worst happen (as it well might: I'm not deluding myself) an invading Russia would probably refrain from nuking a nuke-free Britain. Britain uncontaminated would be a more comfortable and profitable conquest than a radioactive island strewn with rotting corpses. (N.B. I am not a Communist and never have been and never will be. I just believe in adapting to the Nuclear Age — and fast.)

What would happen if "America's support of European defence could no longer be relied on"? Lord Carver regards this as "a totally unrealistic political scenario" but briefly considers it — presuming that Britain has retained her nukes — and concludes: "I do not believe we would be supported by our European allies, who would begin to seek some accommodation with Russia, a very dangerous form of Ostpolitik." He evidently regards seeking some accommodation with Russia as *more* dangerous than our present nuclear balancing-act which, if anything goes wrong, will pitch most of humanity into oblivion and leave the rest to mutate as best they may.

At the end of a conventional war the survivors can somehow pick up the pieces and get going again, though perhaps under unfavourable conditions, if the pieces have been dyed Red. But after the holocaust — that's it, mate. No winners, no losers, no nothing. So how can men of intelligence, kindliness and sensitivity seriously consider the unleashing of a major nuclear war under any circumstances whatsoever? The answer must be that they find it psychologically impossible to entertain the concept of *surrender* — even when the only alternative is extinction. They are still way back in pre-nuke days and at the word "war" their adrenalin begins to flow — you can see it happen — and they want to be away out there fighting for Queen and country and wife and children and home and hearth. At the deepest level they have failed to realise that life's not like that any more — that though appeasement may have been all wrong in the '30s it could lead to the lesser of two disasters in the '80s. Emotionally they are unable to accept what Eisenhower said twenty-five years ago: "We are rapidly getting to the point at which no war can be won." Now we have reached that point, yet too many men cannot adjust to the fact that Science and Technology have suddenly rendered obsolete the immemorially honourable role of "warrior". And so they reinforce IMP by enthusiastically collaborating in war-games.

When I argued for a nuke-free Britain with one very dear friend of mine, an ex-Gurkha officer, he reacted as though I had suggested that he, personally, should be castrated. Yet on most topics he is a man of uncommon intelligence and cool objectivity. Britain, he assured me, needs nukes to defend the Magna Carta and all that. Knowing his genetic content (very Raj) I divined that he also, perhaps unconsciously, feels that Britain needs nukes to prove that the nation's virility has survived the loss of empire. The logic there always eludes me, but I can understand the underlying emotion.

I then asked Roderick the cliché question, which at this crisis in human history is becoming daily less simplistic. "But *of course* I'd prefer to be dead" — and he meant it. Next I asked if he would also choose death for his wife and three adolescent children and he repeated that "Of course!" — but less convincingly, and with a slightly sheepish glance at the door through which his more flexibly minded family might have appeared at any moment. It's one thing for a man to die gloriously on the cliffs of Dover, wielding his kukri against oncoming hordes of Reds — which Roderick would be delighted to do, any day of the week — and something else again for a man's wife and family to be nuked. However, the male mind tends not to dwell for long on these personal matters (that's partly why we're up the nuke creek) and soon Roderick was on to strategy and tactics, and the need to preserve the British way of life. I thought then how lucky I am to be Irish, with no Magna Carta to be defended at all costs.

Afterwards, reflecting on that and several similar conversations, I began to suspect that many men suffer from a surplus of intellect which causes them to elevate Democracy, Freedom,

Tradition, Principles, Institutions etc. above human survival. Splendid as these things are, they're not much use if there's no one around. Mention nuclear disarmament to such men and watch them go floating away in a cloud of high-minded and potentially lethal abstractions. Whereas I, a simple soul of limited intellect, would settle for life on a collective farm rather than have an end put to this splendid experience of being human. Under a Communist régime I would not of course be at liberty to criticise the Soviet equivalent of Lord Carver, if you can conjure up such an entity. But I would still exist, and so would my daughter, and my friends and neighbours; and eventually my grandchildren and great-grandchildren. And by the time those last had been born, if not long before, the Russian invaders that a nuke-free Britain must expect would probably either have left for home or become converts to democracy. However you look at it, humanity *must* do better without the nuclear fix.

It is hard to comprehend the precise nature of this trap into which IMP has driven us — *and* driven itself. Yet we can only escape from it if the ordinary people of every free country oppose their common-sense to IMP's lunacy: which they won't do until they *have* comprehended . . . Meanwhile the British government is trying — happily with decreasing success — to breed a nation of ostriches who will keep their heads in the sand until it's too late to act. This IMP-ishness was exposed during the spontaneous public debate on Civil Defence sparked off by NATO's cruise missile decision in December 1979.

On 21 January 1980, in an article in *The Times* ("The Deterrent Illusion"), Lord Zuckerman, the government's chief scientific adviser from 1964 to 1971, demolished the reasoning behind the idea of a "limited" nuclear strike and wrote: "It is still inevitable that were the military installations rather than cities to become the objectives of nuclear attack, millions, even tens of millions of civilians would be killed . . ."

On 30 January 1980, in a letter to *The Times*, Professor Michael Howard, then Chichele Professor of the History of War and a Fellow of All Souls' College, Oxford, urged the government to "revive" Civil Defence and observed: "The presence of cruise missiles on British soil makes it highly possible that this country would be the target for a series of pre-emptive strikes by Soviet missiles." (Lord Carver, it will be remembered, considers the new missiles *less* likely to draw such an attack than the present air bases.)

On 12 February 1980 *The Times* reported: "Mr James Pawsey asked the Home Secretary if, further to a reply he had given in January on the protection for the public in time of war, he would take steps to advise the public on protection that could be taken now . . ." Mr William Whitelaw, in a written reply, said: "Most houses in this country offer a reasonable degree of protection against radioactive fallout from nuclear explosions and protections can be substantially improved by a series of quite simple do-it-yourself measures." This reply would be laughable in another context; in the context of Britain being nuked it makes one very angry indeed. Yet Mr Whitelaw is neither stupid nor callous; those sections of Northern Ireland's population that most needed compassion, during his term of office there, remember him still with affection. This is one of the most disturbing features of our age; once IMP is at the controls, individual government ministers have to go the way they are driven and the benefits of their personal integrity and common-sense are lost to their country.

On 19 February 1980 Mr James Scott-Hopkins, Euro-MP for Hereford-Worcester, outdid Mr Whitelaw. The *Worcester Evening News* quoted him: "Releasing details to the general public of a Home Office pamphlet, 'Protect and Survive', describing what to do in a nuclear attack, would cause unwarranted panic and be an irresponsible action. With the limited amount of spending money available, Britain should place priority on building up its armed forces." One can easily believe that this pamphlet would cause panic, but with the possibility of "limited nuclear war" being openly discussed, such panic might not be *unwarranted*.

The celebrated Home Office circular No. ES 8/1976, issued on "a limited basis" to the chief executives of local councils, began with this odd statement: "Some of the information in this circular may offend individual beliefs. Recipients may wish to restrict its distribution to those who have a need to know." Immediately one asks: "But do we not *all* have a need to know what our fate is likely to be in the case of a nuclear war? Why treat the public like children who must not be told about frightening things? These, after all, are the people who are footing the nuke bills. Should they not be told what they may get for their money?"

The circular continues: "Due regard has been paid to the present economic climate. Nothing in this circular is to be construed as an invitation to incur expenditure over and above that already authorised for Civil Defence purposes. An additional copy is enclosed for the county emergency planning officer and *further copies are not available*." (Britain was then spending about 40 pence per head per annum on civil defence and more than £200 per head per annum on weapons and military service.)

Next comes the strong meat:

As soon as hostilities intensify, and particularly if nuclear weapons are used to any significant extent, the breakdown of these services, on which most of the public unquestionably rely, would be inevitable over much of the country. Water would not flow from the tap or into the sewerage system. Electricity would be cut off. Refuse collection would cease. Large numbers of casualties would lie where they died. In such conditions, certain diseases could spread rapidly. When radiological conditions permitted movement, district and London borough controllers should assume that one of the priority tasks for their staff, in areas where survivors were to continue residing, would be to collect and cremate or inter human remains in mass graves. If the identity of a corpse is manifest, the burial parties should merely record the name, sex, approximate age group and the place found, for onward transmission in due course to the district or London borough controllers' headquarters. The location of mass graves and the method of disposal would be a local *ad hoc* decision at the time, having regard to the availability of peacetime facilities, the location of bodies, the availability of suitable temporary sites and the importance of avoiding additional contamination of water supplies . . . Once the initial clearance of corpses had been completed, there would still be a problem of several weeks, and perhaps months, of an above average rate of dying from disease and radiation effects. Nevertheless, a return to the pre-attack formalities should be the objective in the longer term . . .

The circular warns:

No part of the country could expect to avoid the effects of an attack. Those areas not directly attacked might suffer from radioactive fallout; would certainly feel the effects of the destruction and disruption elsewhere of supplies, services and transport, and might receive an influx of refugees. All Health Authorities must therefore plan to meet the consequences of an attack on any part of the country. After an attack, the number of casualties might be quite beyond the resources of existing health services. Hospitals might be destroyed or isolated and the care of casualties might have to be undertaken largely by volunteers working in the community under professional supervision. Radiological conditions may be expected to prevent any organised life-saving operations for days or weeks following an attack. Trained health staff would be vital to the future and should not be wasted by allowing them to enter areas of high contamination where casualties would, in any case, have small chance of long-term recovery . . . The Regional or Area Health Director concerned would be responsible for the deployment of ambulances but, in his absence, the highest Director able to exercise control would take this responsibility. Additional vehicles might be obtained by requisition . . . In high casualty areas, however, no arrangements for the deployment of ambulances could deal adequately with the numbers involved . . . Despite any damage and the disruption of public utilities and services, the surviving hospitals might be expected to offer the best facilities for surgical procedures. The Regional or Area Director would have to enforce strict priorities for the admission of casualties to prevent hospitals being overwhelmed. In general, hospitals should, initially, accept only those casualties who, after limited surgical procedures, would be likely to be alive after seven days, with a fair chance of eventual recovery. The more complete the recovery that could be expected the higher the priority for admission. People suffering from radiation sickness only should not be admitted.

One innocent friend of mine deduced from this last sentence that radiation sufferers could recover *without* treatment. However, most people today realise that radiation sickness can be cured — if at all — only by prolonged and intensive treatment which could not possibly be provided after a nuclear attack. So the victims of radiation sickness must be left to die on the streets or amidst the rubble of their homes because they are already doomed. The first symptoms — nausea and vomiting — are usually noticeable within hours. Gradually the vomiting becomes more severe and is accompanied by diarrhoea, weakness and extreme mental depression. The hair and teeth fall out and bleeding starts from the mouth, nose and bowels. It can take several weeks to die. And there are no alleviating medicines.

The ghastly vision conjured up by this circular has nothing in common with Mr Whitelaw's picture of a population snugly tucked away in their "reasonably protected" houses. If widely and persistently publicised it might jerk us out of our present perilous lethargy — hence the government's determination to restrict its circulation. As the bottom line says: "This circular has been issued on a limited basis."

With a stark and horrible logic, man's harnessing of the "basic power of the universe" has brought us close to the point at which we will be unable to behave as normal men and women. Imagine the frenzied grief of a divided family, with some members in "areas of high contamination" and others in comparatively safe areas — the latter knowing that if they went to seek out their horribly dying parents or children, husbands or wives, siblings or friends, they too would die. Could the "safe" survivors remain sane, knowing that the inaccessible people they loved might take weeks to rot away in loneliness and despair, their bodies slowly disintegrating under the influence of the basic power of the universe — harnessed by clever scientists but too strong to be controlled by IMP. As one might harness a fresh young horse, who then bolted . . . Harnessing is only half the job.

If what I have written sounds like morbid fantasy, remember that the governments of many countries are planning for this monstrous thing, this total abandonment of every refinement of the spirit and the heart developed by mankind throughout millenia. The aftermath of war has never been pretty, comfortable or inspiring. But until now man's weapons have not attacked the soul, entailing the abrogation of responsibility towards survivors and the suppression of every decent human feeling — love of family and friends, loyalty to one's neighbourhood, tenderness for the injured, reassurance for the fearful, compassion for the dying. By aspiring to become more than human, through scientific feats, we are now in imminent danger of becoming very much *less* than human.

On 7 August 1945 the late London edition of the *News Chronicle* reported: "In the White House President Truman announced that the first atomic bomb had fallen on a Japanese city." Is it being too fanciful to read some significance into the use of "had fallen on" instead of "had been dropped on"? The latter usage would be the more normal; bombs don't just happen to fall on cities, like handkerchiefs out of pockets. But "had fallen on" implies a certain inevitability, whereas "had been dropped on" suggests some men *deciding* to drop the bomb and other men *dropping* it. Perhaps this slightly odd phraseology expressed an unconscious wish that the bomb had indeed "fallen" by accident out of the blue morning sky — and thus it may have been the very first symptom of Professor Lifton's "nuclear numbing".

When I sat down to begin this chapter, I arrived at a personal understanding of exactly what Professor Lifton, a psychiatrist of Yale University who has spent many years studying the atomic-bomb survivors, meant. To read about the consequences of nuclear warfare is, for many people, merely to gather knowledge about something so evilly final that the more one reads the less one feels. But a writer cannot write *without* feeling about his/her subject. So a thawing of my own degree of nuclear numbing had to take place before I could effectively put pen to paper: and it was as painful as any physical thawing of numbed extremities. What I knew and what I felt had to be "harmonised" and by evening I was close to despair.

How to thaw the nuclear numbing of those who control IMP? This may never be possible, in which case humanity is assuredly on the last lap. Whether the holocaust comes in one, ten or fifty years, come it will if the world does not disarm. But to give in to despair is the cardinal sin. So every "thawed" person must do what he/she can to help thaw others, even at the risk of being considered boring, morbid or cranky. In another context F.S.L. Lyons, until recently Provost of Trinity College, Dublin, has written: "Blaming governments is not only the too easy way out, it is also the sympton of a more deep-seated malaise, which consists rather of the abdication of private responsibility than of the breakdown of public authority."[22] It has now become the private responsibility of every individual to oppose nuclear weapons, since IMP's MADness can be countered *only* through our united efforts. But people cannot be expected to recognise their responsibility until they have somehow been made to think — *and feel* — about the unthinkable.

Many people see surveillance of disarmament as a formidable impediment. Yet Paul Bennett has written: "Our close-look cameras aloft are reaching the ultimate in sharpness and detail. Incredible as it sounds, from a hundred miles up these orbiting cameras can currently spot objects or features as small as one foot in length or smaller. That level of detail allows Pentagon and CIA photo interpreters to locate, identify, and *count* the Soviet's nuclear forces precisely. Missile silos, launch control systems, airbases, bombers on the ground, naval bases and submarines in port are all visible. Factories, submarine construction yards, highways and railroads stand in clear view. There is little chance that the Soviets could build, transport and deploy weapons in excess of SALT ceilings without detection by the US. It would be foolish and dangerous to sign an arms agreement that relied on trust of the Soviets. The SALT II treaty does *not* depend on trust. It relies on sophisticated intelligence-gathering devices which constantly peer inside the Soviet Union to monitor all testing and deployment of nuclear weapons. The UCS has determined that American monitoring systems are capable of alerting us to any violation which could develop into a security threat."[23]

Mr Bennett's complete faith in electronics may seem a little naïve to some people, yet undoubtedly what he has to say is reassuring. So what are we waiting for? As *The Times* remarked, on the occasion of the 508th futile meeting of the UN Disarmament Committee: "Over the years they have worked their way over the whole canvas of the subject and could now well produce watertight treaties at the drop of a hat, if only the political decisions were taken."

Not many people now remember that more than a quarter of a century ago, on 10 May 1955, the Russian delegate to the Disarmament Committee proposed armed manpower ceilings of 1,500,000, massive reductions in conventional armaments and an arrangement for the abolition of 75 per cent of the world's stock of nuclear missiles and other weapons of mass-destruction. He also agreed to an international staff of inspectors who would be given full access at all times to all objects of control. On the completion of these measures, a further reduction of armaments would take place.

The French delegate said: "The whole thing looks too good to be true."

The British delegate, having consulted Whitehall, said: "I'm glad the Western policy of patience has now achieved this welcome dividend, and that the Western proposals have now been largely, and in some cases entirely, adopted by the Soviet Union and made into its own proposals. We have made an advance that I never dreamed possible."

The American delegate, at the end of a two-day discussion with the Pentagon and the White House, said: "We have been gratified to find that the concepts we have put forward over a considerable length of time . . . have been accepted in a large measure by the Soviet Union."

Then, on 6 September, after months of dithering, America killed the world's hopes. "The United States does now place a reservation upon all of its pre-Geneva substantive proposals . . . on these questions in relation to levels of armaments."[24]

Perhaps Eisenhower was remembering this tragic IMP victory when he warned, in the course of his historic farewell speech in January 1961: "We have been compelled to create a permanent armaments industry of vast proportions and this conjunction of an immense military establishment and a large arms industry is new in the American experience. The total influence — economic, political, even spiritual — is felt in every city, every state house, every office of the Federal government . . . In the councils of government we must guard against the acquisition of unwarranted influence, whether sought or unsought, by the military/industrial complex. The potential for the disastrous rise of misplaced power exists and will persist . . . Under the spur of profit potential, powerful lobbies spring up to argue for even larger munitions expenditures. And the web of special interest grows."[25]

The risk of accidental nuclear war also grows. In 1962 Dean Rusk, then US Secretary of State, pointed out: "The danger of outbreak of war by accident . . . grows as modern weapons become more complex, command and control difficulties increase, and the premium is on ever faster reaction."[26] The Pentagon estimates that during the '70s each "early-warning" system gave a false signal about once every three months. As these systems become more various, ingenious and sensitive, their "cleverness" is counterbalanced by their greater vulnerability to mechanical indispositions and human errors. Also, they may be about to enter a science-fiction stage. According to Dr Bernard Knight: "Recent work on micro-chip computers has shown that when a certain stage of complexity is reached the computer begins to have strange properties that are beyond the bounds of theoretical anticipation. The total capacity becomes greater than the sum of the individual parts . . ."[27] Presumably the best brains are in charge of such equipment; but presumably, too, the best brains were in charge of the Teheran hostage rescue effort. At present the morale and training (and much of the equipment) of the US armed forces are notoriously defective. And although we are assured that nothing could go irrevocably wrong with the early-warning systems, because of a multitude of built-in safeguards, IMP's record fails to give their assurances, on any subject, the ring of truth.

Some interesting figures are available for 1972, during which, in America, approximately 120,000 people had access either to nuclear missiles or to the nuke release process. Of these, some 3,650 were dismissed because of instability: mental illness, alcohol addiction, drug abuse or discipline problems. Between 1971 and 1973, 1,247 NATO staff were sacked for one of the same reasons.[28] As all NATO's weapons are directed at Russia, any error made, or any act of lunatic insubordination, would have consequences that no nuclear strategist can foretell precisely. If there is a Plan to deal with such consequences, it has not been revealed.

Nuclear fission is a novelty which imposes unique strains on those responsible for the running of nuclear power stations or the guarding of nuclear weapons. Possibly, seen from the human limitations angle, our harnessing of the basic power of the universe was an evolutionary error which has enabled man to tamper with Nature in a way that offends her — and so must bring on our race extinction. Granted, this is an excessively gloomy if not entirely implausible suggestion: the sort of thing that comes into one's mind only at the end of a longish chapter on nuclear weapons.

In America each Titan silo is guarded by two armed men, capable of launching the missile, who must spend long hours in a confined space. Each has orders to shoot the other if he shows the slightest sign of abnormal behaviour and more than thirty Titan guards have been seriously psychologically disturbed since those missiles were first deployed.[29]

Writing in the *Guardian* on 9 October 1975, Jonathan Steele reported that at one site near Omaha, Nebraska, a young officer said: "We have two tasks. The first is not to let people go off their rockers. That's the negative side. The positive one is to ensure that people act without moral compunction." In a few words, that young man summed up our civilisation's collapse. Suddenly, reading them, IMP appears naked — with no "just war" figleaf — as the instrument of Evil. And one remembers E. P. Thompson's diagnosis: "Deterrence is not a stationary state, it is a degenerative state."[30]

On 16 March 1980, during a Radio 4 discussion on "Home and Civil Defence", the Minister of State for Home Affairs in charge of civil defence, Lord Belstead, said: "If we had the sort of nuclear attack which we think we can expect, about fifteen million people would survive. We reckon that if people take the measures which are being recommended . . . that figure could be doubled . . ."

In other words, between twenty-five and forty million British citizens would be killed in the sort of war envisaged by the nuke-toting British government. And why is that government prepared to fight that war? Nobody seems quite sure. One can only assume it is because of some misshapen offshoot of patriotism which requires most of the population to be Dead rather than Red. But has anybody ever suggested holding a referendum to find out whether the millions whose deaths are being thus calmly contemplated, by their nuclear-numbed leaders, would *themselves* prefer to be Dead rather than Red? If the reality, rather than the Home Office's version, of nuclear war were carefully explained, not forgetting the long-term consequences, is it likely that a majority would choose Death? The British are not the Samurai. And where there is Life there is hope.

8

Yesterday's Men

> It may be assumed that nuclear energy will be harnessed to industrial uses . . . All the existing patterns of life will be disrupted and new patterns will have to be improvised to conform with the non-human fact of atomic power. Procrustes in modern dress, the nuclear scientist will prepare the bed on which mankind must lie; and if mankind doesn't fit — well, that will be just too bad for mankind. There will have to be some stretchings and a bit of amputation — the same sort of stretchings and amputations as have been going on ever since applied science really got into its stride, only this time they will be a good deal more drastic . . .
>
> ALDOUS HUXLEY, 1946[1]

Halfway through my nuke researches, I realised that something was missing. Although I usually write about *people*, there was not one flesh-and-blood pro-nuke in my notebooks — only names, dates, statements, events, opinions, quotations and statistics. It suddenly seemed necessary to listen to pro-nuke arguments being put forward by real live men. The possibility that they might "convert" me was remote, but at least they should be given the chance to try. Soon after I set sail for Britain, to invade the nuclear citadel.

No easy entry was immediately apparent: the worlds of nuclear scientists and travel writers lie far apart. Moreover, I was already known within the citadel as an "enemy"; there had been several hostile letters to the editor of *Blackwood's Magazine* in response to my anti-nuke article. Mr A. E. Souch (Chief Engineer, Planning and Services Division, South of Scotland Electricity Board) commented: "There is an emphasis on the so-called martyrdom of anti-nuclear people which betokens a desire to create a cause whose political expression will be found outside the ballot-box."[2]

This remark neatly illustrated the pro-nukes' "dissident" phobia; anyone fussy enough to fret about the futility of the NPT, or the disposal of high-level waste, or plutonium-armed terrorists, is accused of conspiring to "overthrow society".

Still more revealing was a letter from Major-General S.W. Joslin (one-time Chief Inspector of Nuclear Installations) who began: "I am sorry that you have opened your columns to controversial subjects such as nuclear power."[3] The nuclear industry, being a child of the military establishment, is notoriously furtive. Yet, in a democracy, controversial subjects that profoundly affect the well-being of society, and involve the expenditure of vast sums of public money, are normally considered appropriate subjects for debate in intelligent journals. One recalls Aldous Huxley's warning: "It is probable that all the world's governments will be more or less completely totalitarian even before the harnessing of atomic energy; that they will be totalitarian during and after the harnessing seems almost certain."[4] Huxley got his timing wrong, but this prophecy is now well on the way to being fulfilled.

Happily for me, the Major-General's high-handedness is now officially obsolete. It has been realised that a refusal to communicate merely fuels anti-nuke fires and strenuous efforts are being made to create a new image — all open and smiling and sweetly reasonable. Cheerful PROs answer questions (not always truthfully); films explain how fission works; models show how reactors are designed; glossy (free) information packs have been zealously distributed to counteract anti-nuke literature. And so, despite my record — or because of it? — I was not only welcomed into the citadel but received with fanfares.

This unprecedented foray into the world of Science and Technology caused much mirth among my friends — naturally enough. Changing an electric-light bulb gives me the tremors and after forty-five years of cycling I cannot mend a puncture. So what did I think I was doing, people asked, studying *nuclear science*? Shuffling my feet, I protested that I was not studying nuclear science but nuclear *scientists* — a rather different matter. This, however, was not entirely true. Secretly, I was beginning to enjoy my brush with the atom, after a gruelling initiatory period during which I went to bed every night in a state of severe intellectual prostration. It is perhaps true — as my father used to say, but I never believed him — that all subjects are interesting if taken seriously. This masochistic travail was not of course essential to my purpose. I could never have learned enough to write a technical book, even had I wanted to do so, and my strongest anti-nuke feelings are inspired by issues that have little to do with uranium-enrichment methods or reactor design. Yet my efforts to grasp the basics of the technology seemed worth while when I was at last in the company of nuclear scientists. The minuscule amount of knowledge I had acquired enabled me to sympathise, in a way I could not previously have done, with their excitement and involvement when they spoke of their careers as pioneers, or of the daily difficulties with which they must still cope.

However much one may deplore, fear and oppose the nuclear industry, it does exercise the fascination of all adventure, of every challenge to the human intellect and spirit. Nuclear scientists have to be clever, brave, persistent and — within the confines of their technology — imaginative. These are qualities that have been respected throughout human history and the nuke menace lessens neither their intrinsic value nor their power to evoke admiration. Nuclear scientists are pioneers *par excellence*. They can do things that no previous generation ever dreamed of doing and the difficulties they have overcome are incomparable. Even the astronauts' use of technology is less daring, though their exploits are so much more sensational. It therefore surprised me to find that these unique geniuses were in most respects perfectly ordinary men. And there came a nice twist to that discovery, at the end of one luncheon party at a power station — the six scientists and engineers sitting around the table admitted that they were surprised to find their ostensibly exotic literary guest a perfectly ordinary woman.

Everywhere my pro-nuke hosts were genuinely friendly and hospitable — far beyond the calls of either duty or expediency — and they displayed a quite saintly patience while explaining abstruse technicalities for the benefit of the visiting numskull. Their lucidity was balm to my addled brain; for months I had been immersed in volumes of turgid nuclear jargon and to meet scientists who used the Queen's English had a positively therapeutic effect. Predictably, the more elevated my mentor, the clearer was his language. At Dounreay, whose Director, Mr

Clifford Blumfield, might be described as Britain's nuclear Pope, I astounded myself by being able to follow the theory behind the new process of waste separation now being jointly tried out by the US and the UK. I doubt if anybody but Mr Blumfield could have achieved this didactic feat. (Admittedly, it took him a very long time because of my slow-wittedness, and he used up a great deal of chalk on the blackboard in his office.)

However, old habits die hard and it soon became clear that "industry representatives" would never frankly discuss *controversial* problems with an anti-nuke. It did me no good to point out that most of my readers are British tax-payers who have a legitimate interest in the workings of an industry which they finance and which will endanger them and their children if its problems are not solved. Yet the pro-nukes are not wholly to blame for their repeated ducking behind the Official Secrets Act. This allergy to frankness is partly the fault of certain anti-nuke campaigners who are as expert as anyone in the use of "dirty tricks". Also, it takes more than a change in official policy to change the attitudes of an élite corps. Britain's nuclear industry is staffed by specialists long accustomed to the unquestioning deference of politicians and civil servants. These men have rarely had to discuss their work — or their mistakes — with outsiders. And, after a lifetime of being thus artificially cloistered, it would be unreasonable to expect them to talk shop openly with the hoi polloi. But this era of artificial cloistering may now be drawing to a close.

In February 1980 the Commons Select Committee on Energy uninhibitedly attacked the AEA. Giving evidence to the Committee on 6 February, Mr G. England, Chairman of CEGB, had said: "Regrettably, I cannot produce any other country that has had such an unhappy experience in its nuclear activities over a period of years, in terms of the provision of new plant."

A week later, when discussing reactor choices, the Committee challenged Sir John Hill: "The UKAEA now has a remarkable record of erroneous assumptions and mistakes . . . You made an error on the SGHWR; you changed your mind on that. Next, you changed your mind on the question of PWRs. You were rather emphatic about AGRs . . . Have you learned anything from your past mistakes?" It would seem the answer is "No", since the AEA supports the construction of two new and entirely superfluous AGRs — at Heysham and Torness.

More serious than the nuclear scientists' aloofness is their obtuseness, or nuclear numbing. My new friends not only evaded discussion of the main problems: many clearly *did not believe in them*. A few grudgingly admitted that proliferation is slightly worrying, but most insisted that the NPT is an adequate safeguard. And one affable gentleman had worked out his own solution. "If we" (he meant First World exporters of nuke technology) "if we agreed between us to bomb any Third World installation where weapons were being illegally made, *that* would put a stop to it!" He was not unsuccessfully trying to be funny: he meant exactly what he said. No civil reply seemed possible — and anyway I was there to listen, not to argue. But my stomach felt quite queasy; apart from brandishing a somewhat outdated approach to the conduct of international affairs, this influential figure had just proved himself ignorant of the essentials of the proliferation problem.

A few days later, during a discussion on nukes in the Third World, I questioned the morality of exporting High Technology of any sort to countries that are both unready to cope with it and unable to afford it. My host smiled tolerantly at this reactionary waffle and said, "But what about India? You know India personally — would you not agree that the *only hope* for that country is the spread of nuclear power as rapidly as possible?"

After a longish pause, to recover my breath and recall my statistics, I pointed out that most Indians live in villages of less than 500 inhabitants — and only 11 per cent of those hamlets have electricity — and where it is available only 15 per cent of the population actually use it. But still my host could not see that nuclear reactors are geared to "developed", densely populated, industrial/urban complexes; not to hamlets of a few hundred rupee-less peasants scattered throughout the length and breadth of a subcontinent. No doubt it was naïve of me to be so taken aback by his attitude; High Technologists cannot be expected to view Third World needs realistically. Yet it *is* disconcerting personally to encounter such scientific brilliance co-existing with such a lack of common-sense — which is all one needs to perceive the incongruity of Indian nukes. And this irrational faith in "the nuclear fix" is dangerous not only because of proliferation.

Imported nuclear technology merely fortifies the Third World rich, making it that much more difficult for the poor to get out from under. Also, such programmes do immense harm by absorbing scarce capital needed to develop what have recently become known as "socially benign energy production technologies", i.e. common-sensible projects. These, by partially reducing the use of wood and dung for fuel, would help to check deforestation, soil erosion, soil nutrition loss and the consequent lowering of food production.

Brian Johnson, a Fellow of the International Institute for Environment and Development, has explained how the UN supports nuke imperialism.

In view of the expansion of UN and other multilateral aid and technical assistance programmes, it might be assumed that energy, and especially alternative and where possible indigenous energy sources, would be a priority concern and a principal focus of concessional finance. This has not, however, been the case. Without any doubt the capacity of the UN system in particular to offer a balanced service of advice and technical help . . . has been seriously skewed in favour of High Technology centralised energy systems . . . The only energy agency in the UN system is exclusively concerned with nuclear power . . . [and] . . . past energy development policies . . . have always accepted the primacy of national strategic interests or market forces, which block consideration of small-scale, non-electric, non-commercial energy technologies. The multilateral aid agencies thus concentrate their work largely on the commercial supply of high grade (electrical) energy in large units suitable for industrial purposes. There is an almost symbolic identification of "development" with the hydroelectric dam (or oil-fired or nuclear power station), supplying steel mills and cement factories via expensive high tension electrical delivery systems . . . The reasons for this reflect two conditions which have governed development strategies in the past — and which still constrain them. The first is that simplified assumption of the 1950s, 1960s and early '70s which equated Third World development largely with urban-based industrial expansion, and assumed that advance in this sector was the key to reducing poverty. The second assumption was that energy aid should focus on efforts to encourage centralised production of electricity . . . However, the recent trend of thought on Third World development stresses the critical need to deal directly with poverty, and to respond to the most basic needs for food and shelter and health of the rural majority who are remote from any prospect of urban amenity . . . The place of nuclear power in the strategies of the UN's New Economic Order, which seeks to meet the basic needs of the poorest *directly* by the provision of essential amenities, rather than through the uncertain channels which are supposed to trickle wealth down from urban-based industrial growth, has never been justified. In fact, it has never been officially reviewed.[5]

Only one of my new pro-nuke friends would admit that the threat of nuclear sabotage is real and growing. Yet Flowers (para. 310) takes it very seriously indeed. "Nuclear power stations are exceedingly expensive, costing hundreds of millions of pounds, and as potential targets for a terrorist group would be spectacularly newsworthy. Even if the plant were subsequently repaired, the loss of production could be very large."

This last point was proved in October 1977, when the Hunterston B reactor was accidentally flooded with sea-water; that cost the SSEB £42 million, including repairs and loss of generating power. When I visited Hunterston, a senior engineer admitted that an accident of this type could be caused by sabotage — without the saboteur ever being identified, if he were one of the work-force. I am not suggesting that this particular accident *was* a result of sabotage; its cause — a series of human errors — is clearly explained in the official report on the "event", which will be sent by the SSEB to any interested member of the public. (A welcome glimmer of open government.) But the Hunterston bill does underline the extreme economic vulnerability of nuclear plants.

Discussing sabotage, a lawyer has written: "The possibility of catastrophe arising from malevolent action cannot be excluded. Saboteurs could simply damage the plant, giving rise to real danger of accident, or, if they had the requisite knowledge, they could do so deliberately to cause a catastrophe. The reality of international terrorism is such that it would be idle to deny that terrorist groups could well launch an attack based on good military planning ... Nor can one be convinced that terrorists would not know what to do having captured an installation. Such groups are known to contain highly intelligent and highly trained young people. The work-force at the installation could contain sympathisers, or possibly just persons susceptible to blackmail or threats to themselves or others close to them."[6]

Already terrorists, thieves and practical jokers have operated in power stations. On 4 May 1975 the Baader Meinhof caused two explosions at the Fassenheim power station. On 15 August 1975

Breton nationalists also caused two explosions, at the Mont d'Arree power station in Brittany. And on 14 December 1972, 1,500 workers at Dounreay were sent home after an anonymous telephone call warning of two "suspicious parcels" within the security area. One was found in the main workshop, the other at the entrance to the original Dounreay Fast Reactor. Both proved to be hoaxes, but it chills one to realise that it was *possible* to plant "suspicious parcels" anywhere near an FBR.

Nuclear missile silos and nuclear submarine dockyards are another hazard because, as we have already noted, "eternal vigilance" is incompatible with human nature. And, where nukes are concerned, nothing less is adequate.

On 19 February 1981 a full alert was launched at Britain's top-security Rosyth Dockyard, where the Polaris submarines are maintained. Someone had just discovered the disappearance of a lead-shielded container holding a cobalt-60 isotope used to test the equipment which monitors radioactivity levels near the submarines' nuclear reactors. In a Commons statement, Mr Keith Speed, the Navy Minister, explained that the isotope's radioactivity, while lead-shielded, could not be detected beyond two metres. But if exposed it would emit, within sixteen hours, radiation equivalent to the statutory annual limit for a human being.

On 1 April 1981 (did the thief have a sense of humour?), the empty lead container was found inside the dockyard's radioactivity-controlled zone. At Rosyth the Ministry of Defence said: "No comment." But in London a Ministry spokesman, quoted by the Press Association on 2 April, admitted: "Although exposure to the isotope is not lethal, after twenty-four hours' close contact it would cause local radiation damage to human tissue." He did not explain that the liver is peculiarly vulnerable to cobalt-60, which has a half-life of five years. It emits gamma rays — similar to X-rays and to much of the cosmic radiation received from space. These, in excess, can also cause genetic defects, some of which may remain dormant for generations.

Rosyth-type crime is likely to increase as more people are attracted by the notion of radiation as a personal weapon. In 1979, a Director of the Cracow Institute of General Chemistry tried to murder his ex-wife by planting radioactive waste in her bathroom. And on 31 March 1981, a Cap La Hague employee,

Noel Lecomte, was sentenced to two years' imprisonment for having deliberately used radioactive materials to damage a person's health.

Le Monde noted that Lecomte's trial marked the entry of the French courts into the Atomic Age. The French penal code does not (yet) provide for the use of radiation as a personal weapon; so in accordance with Article 381 Lecomte was charged with "the administration of substances damaging to health". His victim was Guy Busin, a foreman at the plant who, according to the defence counsel, stressed output at the expense of security. This so infuriated Lecomte that he decided to irradiate his boss, hoping that the plant medical authorities would then transfer him to another department. In May or June 1978, he stole three stoppers used to seal magnesium tubes. These tubes hold irradiated uranium rods before they are chopped up, and dissolved in acid baths, to extract their residual uranium and the plutonium created during their working-life. Removing the stoppers from Cap La Hague's high-security area, he placed them under the driver's seat of M. Busin's car — where they were discovered only in March 1979, by which time M. Busin had been exposed to them for a total of some 200 hours. During those months, he felt permanently tired and sometimes dozed off at inappropriate moments. Eventually he had a car crash, from which he escaped unhurt, and subsequently the stoppers were discovered. They were then still giving off more than ten times the radiation dose permitted for nuclear employees. And yet, when M. Busin was examined at the French atomic energy authority's medical centre, the experts could detect no "disability". They were non-committal, too, about M. Busin's future; which makes one wonder why he did not insist upon being tested by independent experts.

According to press reports: "The three stoppers were highly radioactive, but could not cause death."[7] This is yet another example of public befuddlement about radiation effects — a befuddlement fostered by the industry's medical experts and shared, in the Lecomte case, by the judge. That learned gentleman provoked laughter in court when he ordered Lecomte to return the radiated materials, as though they were stolen property of a conventional sort.

Many questions were left unanswered at the end of this trial.

How was it possible to remove those highly radioactive stoppers from the plant in a plastic bag? Why were they not missed? Is there no regular routine check at Cap La Hague on employees leaving the plant, or on irradiated and easily transportable materials? Why did the plant medical authorities not notice M. Busin's condition during those nine or ten months when he was being continually exposed to more than *ten times* the permissible maximum dose? Why did he himself not draw their attention to his increasing debilitation? To what extent does nuclear numbing prevent industry employees from taking what to outsiders would look like elementary precautions?

Back in Britain — on the day before Lecomte was sentenced — three painters died at Rosyth Dockyard after being overcome by what the Ministry of Defence vaguely described as "toxic fumes". These fumes had been released by an unspecified accident inside a dock caisson where men were working. Clearly safety procedures are sloppy at this dockyard, as in so many other provinces of nukedom. And such weaknesses are noted with interest by terrorist groups.

On uranium mining, Britain's pro-nukes can be effortlessly ostrich-like; it all happens *such* a long way away . . . David A. Ogilvie, another of my *Blackwood's* critics, sets the tone: "The nuclear record is an outstandingly safe one. The fuel is intrinsically clean and partially renewable. In its mining, transport and use, the Western world has suffered no major disaster with loss of life in more than thirty years . . ."[8] One wonders how Mr Ogilvie defines the Western world. Between 1946 and 1960, more than 6,000 poor whites and Indian miners in the Southwestern US were "significantly and needlessly exposed to radioactive gases present in the air of uranium mines".[9] Their medical records were closely followed from 1950 and "by 1973 an *excess* of about 180 respiratory malignancies were reported. We estimate that the excess of lung cancer deaths among the total group of some 6,000 miners must be currently of the order of 250–300."[10] To many people this seems just as "major" a disaster as a coal-pit tragedy.

The Navajo and Pueblo Indians have been forced by the US Department of the Interior, working through the Bureau of Indian Affairs, to lease their lands to Continental Oil, Mobil

Oil, Humble Oil, Exxon, Anaconda, Grace, Gold Minerals, Homestake, Hydro Nuclear, Pioneer Nuclear, Western Nuclear, Phillips Petroleum — and of course Kerr McKee, the largest uranium processors in the world, who dealt so ungently with Karen Silkwood when she questioned their methods. On 30 June 1974 there were 380 uranium leases on Indian lands, being worked by grossly underpaid Indian miners.

The Director of Public Health in Shiprock, LaVerne Husen, has specifically condemned the Kerr McKee mines opened in the '50s in New Mexico, the largest centre of the Navajos. "Those mines had a hundred times the radioactivity allowed today. They weren't really mines, just holes and tunnels dug into the cliffs. Inside, these were like radiation chambers, giving off unmeasured and unregulated amounts of radon ... It was a get-rich-quick scheme that took advantage of Navajo miners who didn't know what radiation was or anything about its hazards. Before, lung-cancer among the Navajos was almost unknown, now they're dying of it by the score."[11]

Kerr McKee has of course refused to give any compensation. Their spokesman, Bill Phillips, said to the press: "I couldn't possibly tell you what happened to some small mines on an Indian reservation. We have uranium mining interests all over the world."[12] Each of those "small mines" left behind over seventy acres of radioactive waste when the ore was exhausted; and tailings give off radon gas indefinitely. At present Gulf Oil are sinking two of the world's deepest uranium shafts into the side of Mount Taylor, one of the four sacred mountains of the Navajo and Pueblo tribes who consider it the southern boundary of their universe.

Australia's aborigines have similar problems; much of the country's best uranium lies on their sacred tribal lands. This does not deter the government from planning to mine it, but they have announced no plans for burying the tailings — rather an expensive job. According to dream-time myths, a "rainbow snake" dwells on uranium-rich Mount Brockman and, if disturbed, will bring catastrophe on all mankind. Serve us right, too ...

Arell S. Schurgin and Thomas C. Hollacher have written: "The premature deaths of so many uranium miners must be viewed as a national tragedy and scandal which resulted from the failure of the Federal authorities to set adequately conservative radiation standards, and to require control programmes to achieve and maintain them. The sad fact remains that the present epidemic of lung cancer could have been predicted on the basis of numerous prior investigations in other parts of the world, especially in Europe, where radium mining resulted in analogous epidemics of lung cancer . . . The negligence on the part of government and the mining industry was massive; and may, in part, represent a deliberate disregard of responsibility. Much of the blame for the epidemic must lie with the AEC which should have felt the responsibility to inform mine operators and others of the known radiation hazards . . . Radon and its daughters can be removed from mines by ventilation: the additional cost should be regarded as a normal cost of operating a safe fission technology. We assume that some 500 underground mines will contribute to the production of uranium for commercial power up to 1985. We also assume that the average additional cost per mine over ten years will be $2 million to achieve safe radiation levels. The total additional cost comes to about $1 billion, or 10–20 per cent of the value of the uranium mined."[13]

Perhaps American mining companies will be forced from now on to spend that extra $1 billion; but IMP is not easily brought to heel. In France, for instance, the government permits 600 picocuries per litre of air — *twenty times* the ICRP recommendation and a radon level which caused a five-fold increase in lung-cancer deaths among American miners.[14] The French pro-nukes are nothing if not candid; they have explained that the mining of 80 per cent of the world's uranium reserves would be too costly if the ICRP recommendations were accepted. At their own Brigeaud mine, some twenty million tons of tailings are leaching radium and thorium-230 into the soil and waterways. (Thorium-230 has a half life of 76,000 years.) The mining company admits — again what candour! — that nearby radiation levels are likely to be six times greater than those around nuclear reactors. In 1976 measurements taken on another French tailings pile, beside the Mont d'Ambazac mine, showed radiation levels between twelve and one-hundred-and-twenty times the maximum permitted concentration.[15]

Another slight mining-related snag is the solvent used to extract uranium from its ore in treatment plants. This liquid, which on occasions reaches a radium concentration of 1,000 picocuries per litre, leaches steadily into French soil from its special (but not special enough) storage lagoons.[16] And France is indisputably in the Western world — though of course one has to wait a while for most radiation-induced "loss of life".

British pro-nukes repeatedly claim that one of the chief benefits of nuclear power is an avoidance of mining hazards. Sir John Hill, when Chairman of the UKAEA, wrote: "Mining for coal is obviously a dangerous job and I think we all recognise that it is. On average fifty miners were killed in this country each year over the past seven years, the number reaching sixty-three in 1978. New cases of pneumoconiosis are diagnosed at the rate of 600 per year. In the British nuclear industry we suffer about one fatality every two years. These are mostly caused by conventional accidents but even if half as many again were attributable to radiation that would be one-quarter each year. During this period we produce the equivalent of fifteen million tons of coal."[17]

Here poor Sir John has got himself into rather a muddle. The British nuclear industry produces electricity and plutonium; it does not produce "the equivalent of coal", which in the context would be uranium. Presumably he was trying to say that, during two years, reactors produce as much electricity as coal-fired plants would produce using fifteen million tons of coal. Britain's nuclear industry produces *no uranium*. Its supplies come from illegally operated mines in Namibia, where South African police prevent outsiders from investigating health hazards to the African workers. The average British citizen cannot be expected to know much about uranium mining. But it seems passing strange that the Chairman of the UKAEA should be unaware of the dangers involved.

The saga of Rio-Tinto Zinc's Rossing mine, one of the world's largest uranium mines in Namibia, shows IMP at its most powerful and pernicious. RTZ is a British-based multinational mining corporation which declared profits of £284 million in 1976. In 1975 the *Daily Telegraph* noted: "As well as supplying uranium, copper and other metals, RTZ is also in a position to furnish a coalition government should one be required." This was a reference to the corporation's array of politician-directors, many influential in Foreign Office and Trades Union circles. In 1974 the Conservative Lord Carrington was recruited to the Board and in 1975 Lord Shackleton, a former Labour Foreign Office Minister, became deputy-chairman. Lord Byers, the senior Liberal peer, also adorned the Board, as did Lord Sidney Green, former General Secretary of the National Union of Railwaymen and past President of the Trades Union Congress.

On 24 May 1979 Sir Mark Turner, then Chairman of RTZ, wrote the following letter to Alun Roberts, author of *The Rossing File*:

Dear Mr Roberts,

Thank you for your letter of 18 May, asking me about Lord Carrington. As I am only too well aware, Lord Carrington, on his appointment as Foreign Secretary, is unable to remain a director of this company; indeed, his resignation from our Board was announced a few days after his Cabinet appointment. As an old and valued friend of mine I very much hope that I shall have occasion to see Lord Carrington from time to time, but in his new capacity he is not, of course, available to advise us on any matters. However, we do have a very strong team within the organisation which handles all our international political problems. I believe that Lord Charteris, who was recently elected to our Board and who, in his capacity as Private Secretary to the Queen, has travelled extensively throughout the whole overseas area in which we operate, can also be of great value to us in this particular field.

Yours sincerely, Mark Turner[18]

IMP can seem quite splendidly respectable, especially in Britain: which is why so many people look uncomfortable when attention is drawn to its less savoury manoeuvres.

On 25 May 1977, Mr Alex Lyon, a former Minister of State at the Home Office, stated in *The Times* that relations between the UKAEA and RTZ had become so close that through a "gentleman's agreement" with the government RTZ enjoyed a monopoly of all uranium supplies to Britain.

On 20 December 1975, the *Guardian* reported that Sir Val Duncan, then RTZ's Chairman and chief executive, said the company were "very politically minded, but not party politically minded. If one saw a government was going to do something related to one's business I hope one would know ministers well enough to be able to say so." And on 1 June 1975 the *Sunday Times* quoted a civil servant in the Department of Trade and Industry who was asked about the government's attitude towards securing strategic mineral supplies. He replied: "Oh, if any question of a shortage of anything crops up, you know, we just get on the telephone to RTZ, let them know and leave the rest to them." Now we really are seeing IMP in action.

At present the Rossing mine supplies approximately one-sixth of the Western world's uranium oxide, which sufficiently explains why RTZ, backed by several EEC governments, chose to ignore the UN General Assembly's Decree No. 1. On 13 December 1974 this Decree banned all further mining operations in Namibia and allowed for any mineral resources removed from the territory to be seized and held in trust on behalf of the UN Council. When questioned about the UN ban, at the RTZ Annual General Meeting on 21 May 1975, Sir Val Duncan replied: "I am not prepared to fail to deliver to the UK and others under a contract solemnly entered into for the provision of uranium from South-West Africa. I am therefore not prepared to take any notice of what the UN says about that . . . If that involves disagreement with some of the Resolutions in the UN, I regret that, but that is their problem . . ." To put it more succinctly: RTZ have the bucks and the network, so the forces of international law and order, as represented (feebly) by the UN, can go take a running jump at themselves.

Alun Roberts has pointed out: "Under the terms of Article 25 of the UN Charter, member states are obliged to comply with Security Council decisions even if they vote against them. Therefore, the Security Council's decision of 1971, and the Decree of 1974, affect both RTZ and the British government. Should a future internationally recognised government of Namibia decide to halt uranium exports, or nationalise RTZ, the company would not be entitled to any assistance or relief from the British government. If such measures were taken, the British government, which had persisted with the contracts in spite of rulings by the UN General Assembly, the Security Council and the International Court of Justice, would be powerless to prevent its uranium supplies being terminated."[19] But no Namibian government is likely to oppose RTZ, however many African miners are being abused. If IMP Rule is OK in London, it must surely be OK in Windhoek . . .

The world's reserves of uranium, as of every other fuel, are limited; so the Fast Breeder Reactor (FBR) is a crucial link in that chain with which IMP seeks to shackle us. If the nuclear industry of any country decides not to breed — or is compelled not to, by accidents, costs or public opinion — then in the fairly near future it will become extinct. (At least it has this much in common with humanity.) To continue feeding uranium to thermal reactors would exhaust supplies within thirty or forty years — if not sooner. But the FBR consumes the plutonium produced by thermal reactors while making *more* plutonium from non-fissile uranium, which forms 99.3 per cent of the natural ore but is unusable in non-breeders. This uncanny ability to produce while consuming means that from a given quantity of natural uranium the FBR can extract fifty to sixty times as much energy as a thermal reactor. Britain already has a stock of some 20,000 tons of non-fissile "left-over" uranium, which in FBRs could deliver electricity equivalent to that produced by about 40,000 million tons of coal or 23,000 million tons of oil. (Or so the theory goes.)

According to the AEA: "If Britain implements a programme of building both thermal and fast reactors, then less than half a million tons of crude uranium oxide would be enough to generate all the electricity we need for several hundred years. Without fast reactors an equivalent programme of thermal reactors alone would use up this same amount of uranium every fifteen or twenty years, until a new source of energy became available. Even if this amount could be obtained the competition for the world's limited reserves of uranium could push the price up to levels determined by highly priced fossil fuels . . . Fast reactors thus offer to the world, and particularly to countries like Britain which have to import uranium, both strategic and economic advantages."[20] However, the Nuclear Installations Inspectorate, which issues operating licences, is slightly less enthusiastic: "Given, *as seems reasonably likely,* a successful outcome to the development work now in hand *or envisaged,* there *should be no reason* why a commercial fast reactor cannot be made safe enough to be licensed by the Inspectorate."[21] (My italics.)

Most of the work now "on hand or envisaged" takes place at the Dounreay Nuclear Power Development Establishment, on the wilder shores of Scotland, where it is everybody's ambition to overcome the many complex and dangerous challenges of FBR technology. The nearest town is Thurso and there, in the evenings, the hotel lounge bars unwittingly offer an esoteric extra. For hours one can eavesdrop on small groups of nuclear scientists passionately talking shop — men who plainly live for their work and feel no need ever to get away from it.

In mid-July Thurso was grey, damp, unnaturally cold and pleasingly devoid of tourists. A sullen sea heaved restlessly in the small harbour and John o'Groats was skulking in cloud. The wide surrounding farmlands were flat, windswept, treeless, colourless: like a shred of Peruvian puna gone astray. Exploring the coast with chattering teeth, I wondered what could possibly have attracted Neolithic man to Caithness and why it had remained so popular — with the peoples of the Bronze Age and the Iron Age, and then with the Picts and the Vikings. But that was merely a passing mood, after a long, cold night on a slow train.

The friendly natives soon thawed me; mysteriously, they all had *Southern* Irish accents — a great relief after my abortive attempts to converse with the Glaswegians. They were solidly pro-nuke: Caithness may have a few anti-nukes hidden in the crevices but I couldn't find one. A considerable percentage of the population should be under constant surveillance, Dounreay being one of Britain's most "sensitive" nuclear installations, but nobody seemed to be grieving over diminished civil liberties. Nor did anybody show the slightest interest in other nuclear problems; an anti-nuke demo at Dounreay would probably provoke civil war.

Since 1955, when the AEA's first construction workers arrived in Caithness, the town's population has risen from 3,000 to 10,000 and 1,750 new houses have been built — 1,000 by the AEA. A beaming girl from the tiny tourist office, delighted to find what she assumed to be a tourist, took me to see the new primary schools, the new high school, the new technical college: all offshoots of Dounreay. Later, in a pub, an elderly man proudly informed me that in his lifetime Thurso had been transformed from a stagnant village to a thriving town with a choice of social, artistic and sporting activities. Even more important, many young people from all over the Highlands can now attend one of Dounreay's training schemes — Junior Scientific, Clerical and Secretarial — instead of migrating south. And the Dounreay Apprenticeship Scheme, for scientifically talented youngsters, is recognised as one of the best in Britain. Thurso is indeed an AEA triumph, an arresting example of the nuclear lion lying down with the public lamb — or of ignorance being bliss.

At 9 a.m. a Dounreay car collected me from my hotel; I had suggested travelling from Thurso in a workers' bus but this idea was firmly squashed — Mr Blumfield likes to treat his guests well. A few months later, when the newspapers reported plutonium-contaminated workers' buses at Dounreay, I felt glad that my egalitarian instincts had been thwarted. According to a trade-union official, twelve buses were contaminated — and these are also used to ferry schoolchildren. The AEA admitted to only one contaminated vehicle and did not comment, to my knowledge, either on the schoolchildren allegation or on a *Daily Mail* report that one seat was too contaminated for future use and had to be "disposed of".

As the media are notoriously inexact on nuke details, I wrote to Mr Blumfield who, with his usual courtesy, replied in detail: "No seats have been destroyed. The bus situation arose because some-one did not carry out the laid-down self-monitoring procedure before he left site. We have special monitors which check em-ployees as they leave their areas. These are very sensitive but cannot be quite as sensitive as the equipment which the individual should check with. As a further check we routinely monitor the buses. It is very seldom that anything is found — about once per year. The amount of radioactivity is very low, because our monitoring equipment for employees leaving their area of work is sensitive and would send the individual back for further checks. Our bus monitoring instruments are even more sensitive. If a small spot is found it is removed by us. It would give a radiation dose of about 1/50 of natural background if someone sat there continuously for twenty-four hours every day of the year — or equal to the radiation dose received in five hours when flying in a modern aircraft. We are pursuing the personal monitoring procedures still further and I hope to eliminate the problem completely. You make the point that we are responsible in this country. The procedures we use to safeguard our employees and the public are very extensive. Most of them are self-imposed. For instance, the bus monitoring and extra checks of employees are carried out because I ordered my staff to do it — no one told me to do so. Some people outside the industry have commented, 'If you didn't do it no one would be the wiser.' That is not my idea of being responsible."

What an amount of suffering and sorrow might have been averted if Blumfield-types were common in nukedom!

During the ten-mile drive from Thurso, along a bleak unpeopled coast, the chatty Yorkshire driver explained how easily outsiders settle into Caithness, how good the local schools are and how gratifying it is to work at Dounreay where the *future* is being sorted out. This was not, I felt, mere Public Relations prattle. Although I saw little of the Dounreay work-force, and there was a minor strike of eight men on at the time, I sensed an unusual *esprit de corps* within the Establishment — perhaps inspired by a daily sharing in the unadmitted but unique dangers associated with the operation of an FBR. Not to mention the various other perilous experiments undertaken at Dounreay, like the manufacturing, reprocessing, examining and testing of breeder fuels. And the study of liquid metal as a coolant, a process which itself breeds a shoal of novel conundrums — though Mr Blumfield informed me, with justifiable pride, that most of these have by now been solved. (He did not state whether the solutions found could safely be applied within a series of commercial FBRs; it might be unreasonable to expect the Dounreay level of skill, knowledge and watchfulness among the staffs of electricity boards.) Then there is the development and testing of special instruments and components — one more delicate and complex than another — for use with liquid sodium. And also the development and testing of monitoring equipment for a FBR, this being a particularly taxing problem because the coolant is opaque. But again Dounreay has found the answer: long metal rods known as "wave-guides" which listen to the behaviour of components below the sodium surface in the reactor core. Sound-waves travel up these rods and are detected at the top by sensitive pick-up devices. If these electrical signals are amplified they can be heard over a loudspeaker; or they may be fed into instruments capable of complicated analyses. These techniques are also used for fault-detecting and problem-solving — and the problems of a Prototype FBR are naturally legion.

The *nature* of breeder fuel constitutes one problem unlikely ever to be solved. It has to be far more radioactive than thermal oxide fuel and its high percentage of plutonium greatly increases the risk of criticality. In a thermal reactor a meltdown is the worst that could happen: the fuel could *never* explode. But in a fast reactor melted fuel could "give rise to a nuclear excursion", i.e. blow up. Moreover, the handling (or not handling) of spent breeder fuel is technically as challenging as the safe operating of the reactor itself. At Dounreay I stood in awe outside the row of remote-handling caves — empty of fuel that day — and marvelled at the skill of men who, with twelve-feet-long artificial arms, can perform the most delicate operations on fiercely radioactive spent fuel elements. (I tried to manipulate those Space Age limbs, with no success.)

After their removal from the core, these elements must spend three weeks or so in special containers next to the reactor while the deadliest of their residual activity is being dissipated. At this stage any breakdown in the coolant system could cause a disaster at least as serious as an "excursion" within the reactor. (I'm becoming quite addicted to nuke jargon; all this talk of "excursions" makes me visualise lots of happy little neutrons and protons scrambling on to a Bank Holiday bus with picnic-baskets.) The spent elements are next immersed either in liquid sodium or in a sodium-potassium alloy — which itself becomes highly radioactive, thus augmenting the world's waste-disposal worries. Any loss of coolant would still be calamitous, as would any contact with air or water. The Dounreay explosion on 10 May 1977, when the concrete lid was blown off a waste pit, was caused by the reaction of water and sodium. But, as Mr Blumfield explained in another letter to me: "The sodium explosion was a minor one. Someone made an error when carrying out a process to ensure that an item had no sodium inside it. The item was then stored under water. Later the water entered the item and reacted with the sodium, a well-known process. The procedure was set up to eliminate the sodium to prevent this. The method of storage has been changed so that there is no water there and such a reaction cannot occur."

The reprocessing of FBR fuel, on which the success of the programme ultimately depends, is still at an experimental stage. At present many pro-nukes, who are unconnected with the programme, doubt if reprocessing at an industrial level can be done quickly enough for breeding purposes without exposing workers to unacceptable risks. The snag is that Fast Breeders do not breed fast. They derive their name — and much of their hazardousness — from the fact that they use the *natural* speed of neutrons (10,000 miles per second) which has to be slowed down in thermal reactors. Their actual production of plutonium is slow and is usually calculated in terms of "doubling time". A thirty-years doubling-time means that at the end of that period enough plutonium has been bred to fuel two reactors. However, the doubling-time depends not only on the rate of breeding but on the speed with which reprocessed fuel can be returned to the reactor and on the percentage lost during reprocessing and fuel fabrication. For instance, if in one year an FBR breeds 165 kilograms of plutonium in a total of 2.5 tons, a loss of 6.5 per cent when the total is taken from the core for reprocessing would cancel the breeding gain. And such losses are common; Windscale estimates its losses during reprocessing at more than 7.5 per cent. (Hence the present sad state of the Irish Sea, into which some 266 kilograms of plutonium were discharged between 1971 and 1976.) The spread of FBRs would in fact send the world up a very polluted creek. Which is one reason why Flowers (para. 510) suggested that: "Even if an alternative strategy could not give the same assurance that future energy demands could be satisfied, there is the possibility that the nation would prefer to accept that risk in order to avoid what might be seen as the greater risk of a much expanded nuclear future."

At Dounreay, one can well understand the hypnotic fascination of nuclear challenges for nuclear scientists. It is a strange and curiously moving experience to stand on the "lid" of the concrete reactor vessel, which itself is underground, and know that just a few yards below one's boots the most dangerous process in the world is occurring — and that yet one is safe, because of the extraordinary range and inventiveness of the human mind.

As the day went on, I found myself being quite carried away by the romance of it all. Hearing a few of the subtly baffling difficulties of FBR technology being discussed by a physicist is like watching some exceedingly sophisticated drawing-room game being played by the abnormally gifted. Or one can see FBR scientists as latter-day Marco Polos crossing unmapped continents, never knowing what new peril the next day or the next mile may hold and relying for survival on their own initiative and swift responses. As someone with a certain predilection for risks, I could scarcely have failed to enter into the spirit of the thing at Dounreay.

However, when I was back in Thurso and my effervescence had subsided I sadly faced reality. Then Dounreay looked like an adventure playground for the more daring and creative spirits within nukedom, whose playing with fission could one day bring disaster on the world — the nuclear equivalent of the Seveso accident, as Dr J. Ravetz, of the Council for Science and Society, has pointed out.[22]

Many people oppose the commercial FBR programme because of the variety of risks implicit in a plutonium society — political, environmental, economic and sociological. But Dr Ravetz argues that: "Here the physical hazard is the prime focus of concern. A nuclear reactor is a completely different sort of system from an organism, human or social. Its design has been *conceived* abstractly, and has not had a chance to *evolve* through long experience. It is *brittle* in response to disturbance, not homeostatic like an organism. Accidents not anticipated in the original design are liable not to be buffered, but to rip through the system to culminate in disaster. The external effects of an FBR disaster . . . would be a major assault on our environment and social fabric, in some ways worse than conventional war. In the assessment of technological risks, the calculation assumes the comprehension of the original safety design, and the effectiveness of human monitoring through the whole fuel cycle. All official estimates of probabilities of accidents and their consequences are only *technological forecasts, not scientific facts.* The threat of disaster from the installations is ever-present. Protection depends entirely on the skill of the original designers and the vigilance of safety staff."[23]

If commercial FBRs proliferate, "a major assault on our environment and social fabric" would seem to be inevitable. It is thirty-four years since Aldous Huxley referred to "the non-human fact of atomic power" and, because nuclear scientists are so ingenious and courageous, and apparently capable of competently controlling nuclear fission, we tend to forget that the creation of plutonium puts humanity in a situation *for which evolution has not equipped us.*

Even at Dounreay, where the standards of vigilance are certainly very much higher than at most nuclear installations, two fuel-rods, containing between them thirty-five grams (about 1oz) of plutonium, are known to have been lost — one in 1973 and the other in 1977. Mr Blumfield told the press that these had "almost certainly" gone either to a waste pit or to the recovery area. He added, when asked if the two rods could have been "diverted": "We have methods of checking this sort of thing which I am satisfied safeguards this material." Subsequently he explained in a letter to me: "Plutonium in our fuel pins is mixed with uranium and is in ceramic form. That is, it cannot be burnt, is hard and highly resistant to abrasion, can only be dissolved with great difficulty unless one has sophisticated equipment and therefore would be difficult to distribute. As you rightly put it, an irradiated rod is self-protecting. An unirradiated one can be handled without problems but to change its form as I have said is very difficult. Extraction of plutonium requires not only the technical knowledge but also the equipment. It cannot be done on the kitchen table and an individual would have to be well-trained to achieve the separation. I cannot imagine there being a black market for such amounts and I doubt if someone who has reprocessing capabilities would have an interest in small amounts . . . It is not possible to discuss the control methods but of course control depends on many methods. It is because of these that there is certainty the material has not been stolen . . . 'Materials Unaccounted For' (MUF) is a peculiar term. It relates to the accuracy of the many measurements involved. However precise these are, there is still a margin of error. That is, the accuracy of measurement is not absolute. The manager of any plant wishes to know his material balance for efficiency, financial and control purposes. There is also the accuracy of calculation of burn-up and production of plutonium. When one takes all these into account there is a range of accuracy. It is the imbalance due to this which is called 'material unaccounted for'. It does not mean the material is missing. Overall the calculations are complicated . . . I cannot directly compare our system with those used by the Americans. The procedures obviously depend upon the material and process used. However, some of the equipment we have developed for plutonium control is very advanced and a number of countries are interested." Mr Blumfield made that last observation in response to a remark of mine: "Presumably if the AEA had a safer accounting system than the NRC they would have shared it with their American colleagues for the sake of global security."

At this stage one recalls Israel's "diversion" of hundreds of kilograms of enriched uranium from the NUMEC plant — and also the ERDA statement made a few years ago: "The aggregate MUF from the three US diffusion plants alone is expressible in tons. No one knows where it is, none of it may have been stolen — but the balances don't close. You could divert from any plant in the world and never be detected . . . The statistical thief learns the sensitivity of the system, operates within it and is never detected."[24]

When discussing FBRs the Americans refer to HCDA (Hypothetical Core Disruptive Accident) which could happen if the fuel so rearranged itself as to cause a mini-nuclear explosion ("Mini" of course being a relative term . . .). Such a misfortune might be set off by loss of sodium flow (perhaps caused by pump failure), together with a malfunction of the reactor shut-down systems. Partial meltdowns have already occurred in experimental FBRs, so we are not now just having nightmares.

The AEA depicts FBRs as essential for the future prosperity of the industrialised world; so an altogether disproportionate percentage of r & d funds has been devoted to this technology. However, no full-scale commercial FBR has yet been operated anywhere; and almost unbelievably expensive disasters litter the route of FBR development, like skeletons by a desert track. Many eminent physicists now regard the technology as so intrinsically unsafe that it should at once be abandoned. Apart from *five tons* of plutonium in the core, a commercial fast reactor, such as that now being built at Creys-Malville in France, would contain 6,000 tons of liquid sodium coolant; and, in the scientific world, it is widely believed to be impossible to quench a sodium fire involving more than a few hundred kilograms. The French Super Phoenix is *designed* to withstand an explosion of up to 200 kilograms equivalent of TNT, but nobody knows whether it *could* withstand it. The Centre National de la Recherche Scientifique (Commission 06) has reported: "Certain critical elements such as the reactor vessels of Framatome or the vaulting of Super Phoenix have not been designed on the basis of safety coefficients but purely on the basis of what is technically feasible."[25]

The frightening thing is that here we may have another Concorde-type débâcle — a project generally acknowledged to be doomed to failure but secure from outside interference because it has developed its own momentum.

Let George L. Weil have the last word on FBRs. Dr Weil should know the form, having worked with Enrico Fermi on the Manhattan Project, served the AEC as chief of the Reactor Branch in the Division of Research, and then become assistant director of Reactor Development within the AEC. In 1973 Dr Weil remarked on the AEC's verbal hypersensitivity when discussing FBRs: "No matter how it is phrased — 'nuclear explosive energy', 'rapid reassembly of the fuel into a supercritical configuration and a destructive nuclear excursion', 'rapid core meltdown followed by compaction into a supercritical mass', or 'compaction of the fuel into a more reactive configuration resulting in a disruptive energy release' — the meaning is clear. FBRs are subject to 'superprompt critical conditions' and, as the AEC well knows, this technical terminology translated into layman's language is 'an atomic bomb'."[26]

On the train to Inverness I talked with a young Londoner who had been working at Dounreay for three years but refused to take radiation seriously. To him it was a figment of some bureaucrat's imagination — a nuisance that had given rise to yet another set of tiresome regulations. He admitted that he had never read *Working with Radiation*. This excellent booklet, produced by BNFL for the benefit of their employees, concisely explains radiation risks and the means that should be taken to avoid them. However, I had already observed, around other nuclear installations, that it is not widely read by workers — many of whom are allergic to printed matter, especially if the style is slightly academic. Such literature is reinforced by verbal exhortations, but workers not personally committed to their own safety must be extremely difficult to "nanny" day after day.

Another problem is the bonus temptation. Workers are paid "discomfort money" for wearing protective clothing in radio-active areas and when their radiation badges show that they have received maximum permitted doses they are moved to "normal" areas. The Director of one power station admitted to me that fiddling badges and dosemeters is easy — and commonplace. The film can be removed from the badge or the dosemeter can be wrapped up to reduce the radiation registered. If workers prefer money now to health later, there is nothing much their super-visors can do about it — or so the argument goes. But how many workers share the view of my friend on the train? I suggested to the Director that more detailed instruction on the long-term effects of radiation might help; but he obviously felt personally reluctant to dwell on those effects and so was not disposed to organise radiation pep-talks for his staff. Nuclear numbing has many manifestations.

In Britain, dastardly deeds are no doubt done, behind the impene-trable screen of the OSA; and cover-ups are endemic throughout nukedom; and AEA propaganda is almost as tedentious as its AEC sire. Yet to exchange America's nuke scene for Britain's is like walking from a foetid slum into a rose-garden. As Flowers (para. 175) remarked: "The contacts we have had with the nuclear industry during our study leave us in no doubt that the most diligent attention is given to safety in the design, construc-tion and operation of reactors."

But what about the many important problems that even the most diligent scientists cannot, apparently, solve? This point emerges from a paragraph in the "Findings of the Board of Inquiry into the Ingress of Sea Water into the Second Reactor at Hunterston 'B' Nuclear Power Station in October 1977":

"Warning indications commonly called 'alarms' were then received of 'high fluid levels' in the circulator housing; and also some, but not all, of the reactor gas moisture meters indicated high gas moisture content. The latter instruments, *while the best available,* were known to give varying readings with changes of pressure, temperature and with length of time in service, and were therefore *difficult to interpret.* In particular the characteristic of the meters is such that at very low moisture levels they give a 'high moisture' reading.(! ! !) This causes the warning indication to persist and thus *prevents a correct warning being given* with a change from very low to very high moisture levels." (My italics and exclamation marks.)

These Alice-in-Wonderland meters, which say high when they mean low, do not — I have been told — endanger the public. But they did contribute significantly to the SSEB's loss of £42 million.

The British nuclear industry specialises in record-breaking losses. At the end of March 1976, Professor David Henderson of University College, London calculated that its AGR programme would send some £2,100 million down the drain; he was reckoning at 1975 prices and taking into account predictable future costs.[27] A feature of that particular programme is of course *un*-predictable costs . . . Yet its humiliating story always has been — and still is — one of chronic labour-strife, AEA incompetence or obstinacy and inter-engineering company rivalry: it is not an American-style saga of widespread corruption and cynical irres-ponsibility. Not that this is much consolation to the British tax-payer, especially as the industry continues to shun reality. Its plan to order one new nuke every year during the next decade was described as "over-ambitious" in February 1981 — by the Commons Select Committee on Energy.

This programme would cost £15 *billion* and the Committee's Report criticised the government for favouring nuclear power at the expense of coal-fired stations and energy conservation. It stated that neither the Department of Energy nor the CEGB had been able to justify such an expansion of the nuclear industry and it accused the CEGB of giving "misleading" information about future electricity demand. "The programme announced in Dec-ember 1979 was based on the forecast that peak winter demand in 1986–7 would be 52,000 million watts. Yet within a matter of weeks that forecast was reduced by 7 per cent . . . It would have been less misleading and more helpful if the CEGB had informed us . . . that forecasts had already been overtaken by events and were in the process of being revised downwards, even if the precise figures may not have been known at that stage. The credibility of much of the CEGB's evidence was undermined by this omission and we trust that this will not occur in the future."[28]

The Department of Energy also roused the Committee's ire: "We were dismayed to find that, seven years after the first major oil price increases, the Department of Energy has no clear idea of whether investing £1,300 million in a single nuclear power plant is as cost-effective as spending a similar amount to promote energy conservation."[29]

The Committee concluded that: "Perhaps the most serious indictment of the industry was that Britain pioneered the use of nuclear power, but now had to license foreign technology because its own reactor design remained unproven after sixteen years. Its mistakes had imposed a wholly avoidable financial burden on electricity consumers and tax-payers."[30]

Reading all this, I rather perversely found myself wanting to defend Britain's much-despised nuke industry. Despite their numerous money-gobbling faults, the AGRs are as *safe* as any reactor could well be. No foreign country has ever wanted to buy one — naturally enough, though in conversation with British nuclear engineers this is a subject upon which it would be tactless to dwell. Yet when they *do* work the AGRs are trustworthy, if not efficient. And the same may be said for the original British Magnox reactors, two of which were exported — to Japan and Italy — in the days before the Americans secured their grip on the world reactor trade. The senior Magnox reactor at Calder Hall — grandfather of all Britain's commercial reactors, though it started life in the arms business — is still quietly working away, causing no anxiety to its operators, at the remarkable age of twenty-six.

Some American readers may indignantly contest my comparatively sanguine view of British nukedom. How can I be *sure* that its safety measures, so often exposed as shockingly inadequate, are on balance less irresponsibly implemented than the NCR's? This is where the OSA bites off the British nose to spite the British face; because of it, I cannot prove my contention and must confess that my judgement here is partly based on intuition (of all untechnological things!).

To compile a British equivalent of "The Nugget File" would be something akin to high treason and many more "incidents" must occur than ever reach the public ear. After all, Dounreay's interesting examples of MUF were revealed only in 1980, by an AEA employee who had to remain anonymous. The need for a British Freedom of Information Act is urgent, though one recognises the formidable practical difficulties in the way. In December 1979, David Pearce and Lynne Edwards of Aberdeen University, and Geoffrey Bennet of Sussex University, submitted a report to the Social Science Research Council in which they warned that if some such act is not soon introduced the public will be unable to participate responsibly in the nuclear debate — and therefore debate may be replaced by serious civil disobedience.

Several senior officials of the British safety and regulatory bodies impressed me as men of integrity and dedication, who have set the highest standards for their juniors. But is it likely that such standards can be upheld indefinitely? Some old-timers feel that the necessary level of vigilance may prove impossible to sustain. For younger men, the maintenance, inspection and operation of thermal nuclear reactors and power stations, and enrichment and reprocessing facilities, are, inevitably, routine matters. They have come to a relatively mature technology and have never been creatively involved in setting it up, determining and guarding against its hazards, and experimenting (or closely following the experiments of others) in the fields of physics, chemistry or engineering. To them the industry is just another job-source, whereas the pioneers are every instant aware that the safe operation of a nuclear reactor must always remain a unique challenge. True, the CEGB maintains its own nuclear engineering and research division, runs refresher courses at the operators' training centre and in 1980 decided to provide simulators for each AGR station. But what we are discussing has little enough to do with training or simulators: it is an attitude of mind. To illustrate this "younger" attitude, a senior engineer complained to me that some operators now regard it as a point of honour to keep their reactor going even if something seems slightly amiss, instead of letting the automatic safety system take over. To err on the side of caution, by instantly scramming a reactor without real need, could be an expensive error; yet less expensive than erring in the other direction.

The CEGB, which operates most of Britain's nuclear reactors, is not, mercifully, as self-satisfied as its publicity literature makes it seem. Post-TMI, it didn't simply explain that such an accident would be impossible in Britain because of the differences between gas-cooled and pressurized water reactors. Something more was needed, since the Board is considering the building of a PWR if the NII agrees. It therefore produced a brief but workman-like document outlining the lessons it learned from the TMI disaster and implying that a British PWR will be altogether more reliable than its American progenitor. (A point furiously disputed by many British nuclear engineers.) This document, "The Kemeny Report and the CEGB Response", would, if widely circulated, do more to soothe public unease than the half-truths pushed by the Electricity Council in such pamphlets as "The Need For Nuclear Power". Because that particular pamphlet is aimed at school-children, its clever dishonesty is peculiarly repugnant.

It is interesting to compare the British and American regulatory systems. (The OSA does not prevent this; it only prevents check-ups on the extent to which those systems are taken seriously.) As we saw in Chapter One, America suffers acutely from ABCD. Happily, this disease is unknown in Britain, where there are only two nuke-using nationalised generating boards, one licensing authority (the NII), and — now — one major nuclear plant contractor, the Nuclear Power Company (NPC). Also the Generating Boards and the NII have always worked on the assumption that for the licensee to produce a safety case, to be scrutinised by the NII, is more sensible than for the NII to impose a multitude of minutely detailed regulations on the Boards. Granted, Flowers (para. 287) was unhappy about the NII's working methods: "There seems to be rather little contact between the Inspectorate and the designers during the design stage. The Inspectorate is not formally involved until the complete design has been prepared and submitted to them. The design is analysed in considerable detail and the Inspectorate then indicate any requirements they may have for changes as a condition of licensing. The view has been expressed to us, and we find it convincing, that it is impossible for an inspecting body adequately to ensure safety by analysing the completed design of a large, complex plant. Safety is built into the design as it progresses and the primary task of the inspecting authority should be to ensure that the proper analysis techniques and disciplines are used at every stage of the design process to achieve clearly defined safety objectives." This lack of "contact" during the design stage contributed to the AGR shambles. Yet whatever the defects of the NII-NPC-Generating Boards relationship, it certainly produces safer (though far more expensive) reactors than the American system.

The Kemeny Report recommended that each nuclear plant company should have a separate health-and-safety "watchdog" group to report directly to top management — a precaution adopted in Britain twenty years ago. On the even more crucial matter of operator selection and training, it is inconceivable that Britain's generating Boards would ever have tolerated the American system. As shift-charge engineers they employ graduate, chartered or qualified engineers. An operations engineer, after appropriate training, will work as a control-room desk operator. He may then be transferred to refuelling operations or auxiliary plant, to gain experience of equipment operation, and he is then promoted to assistant shift-charge engineer for a year or more before becoming shift-charge engineer. Meanwhile he will have been attending courses at the Nuclear Training Centre at Oldbury to deepen his theoretical knowledge of reactor plant design, safety principles and fault conditions. Poor Craig Faust wouldn't have got far in Britain's nuclear industry.

Yet one does wonder how even the best-trained and most highly qualified operator would react to a major crisis. When every possible precaution has been taken, the human temperament remains unpredictable. And even nuclear scientists can be sensitive to the permanently *odd* atmosphere of a control-room. My guides all had trouble dragging me away from these eerie places with their magical arrays of dials, discs, lights, bells, knobs, graphs, buttons and charts — at once disturbing and fascinating. Instinctively one lowers one's voice and the first I visited instantly and overwhelmingly recalled the almost inaccessible inner sanctum of a colossal Mayan temple half-buried amidst the dense jungle of Honduras. In these two shrines — apparently so dissimilar — I experienced exactly the same sort of frisson. Which no doubt was very illogical of me. Or was it?

My more extreme anti-nuke friends were shocked to hear that I had found the "enemy" camp thoroughly exhilarating. I tried to explain that at Dounreay, especially, it was possible to appreciate how nuclear scientists *feel*; obviously the struggle to overcome nuclear hazards is as exciting and rewarding for them as writing books is for me. But the extremists could view pro-nukes only with contempt, as essential components of the IMP machine. This of course they are, yet one cannot fairly identify all (or even most) nuclear power scientists with those who manipulate both them and us. Many of these men seem the victims rather than the allies of IMP. Success in the world of Science and Technology demands sacrifices; thus they have never been free to voyage far enough from their specialist island to gather the knowledge and experience that would enable them independently to evaluate industry propaganda. Also, as they themselves regularly solve the most formidable problems, they see no reason why equally competent experts in other fields should not be able to cope with nuclear terrorism or proliferation. And, by some extraordinary feat of mental gymnastics, the majority have convinced themselves that there is no longer any link between their governments' nuclear power and nuclear weapons programmes.

This last exercise is essential because most pro-nukes are as concerned about the future of mankind as any anti-nuke. They are not, as their extremist opponents would have us believe, callously pursuing profitable careers with no thought for the well-being of humanity in general. (That attitude is found in the dark corners of the industry, where the Bosses lurk — and where radiological findings are suppressed, accounts cooked, politicians bought and unprotected workers sent down uranium mines.) The average nuclear scientist is blind to the social/ethical implications of his life-work — as are many other contemporary scientists and technologists. This weakness makes his activities no less dangerous, but it needs to be taken into account if further polarisation is to be avoided. A rising tide of mutual bitterness will do nothing to help resolve the nuke controversy.

A curious — and sad — version of the generation gap is occasionally to be found within nukedom. Some elderly experts, who received a wider education than is now possible for aspiring technocrats, are deeply disturbed by the threats implicit in a "plutonium society". After a few hours' conversation, it is possible to sense their inner unsureness about that nuclear future to which they have devoted their whole lives and considerable talents. One sympathises with such men. It must be distressing, at the end of an exceptionally demanding career, to be forced to wonder if one's best intentions have paved the way to catastrophe.

The argument for nuclear power is invariably obscured by a fog of statistics, predictions, diagrams, estimates, charts, extrapolations, graphs, forecasts, calculations, feasibility studies — and a few extra statistics. But if we ignore the fog and focus on the foundations we find an assumption that man *must* continue to live the way he lives now — only more so. Pro-nukes cannot see that to do this will prove ecologically impossible, leaving aside other considerations. It may be that future generations — if there are any — will see the great nuclear adventure as the climax to an era. We cannot guess how the transition to the next era will take place. Only the *inevitability* of change is certain, because our planet's limitations dictate it. We are not, after all, the Lords of the Universe.

Now my friends are asking: "So what are you telling us to *do*, at the end of your book? How do we go about *stopping* nukes?" Happily, not being a leader, planner, organiser, crusader, ideologist or orator, I feel no obligation to tell anybody to do anything. My value as an anti-nuke is strictly limited to communicating on paper the reasons why we should do *something*. Others must advise *what* . . . For instance, in the January 1981 issue of *Resurgence*, Ronald Higgins suggests some legal and respectable weapons which all citizens of a democracy can, and should, use against IMP.

Lloyd Geering has lucidly explained why it is no mere eccentric hobby for our generation to do this "something":

Modern man sees himself as an indivisible part of an intricate, developing, ecological planetary system . . . and is concerned with the welfare and destiny of the whole man, both physical and spiritual . . . This concern is manifested in such things as the conservationist movement, the need to husband non-renewable resources, population-control, and the accommodation of human practices to the process of nature . . . Men do not want to be known by labels, such as Christian, humanist, agnostic,

to which they do not want to conform. They want to be known for
themselves; they want to be free to be themselves, free to change and
grow to greater maturity . . . Modern man, in order to retain his humanity, and if possible to experience his human potential even more fully,
must make his response to ultimate issues; and that is what it means to be
religious. The ultimate issues which confront him emerge from *within*
this world and sometimes even from within himself, because of the
greatly increased responsibilities placed upon him in the exercise of his
greatly increased freedom . . . We can no longer appeal to God up there,
or to former authorities. We are what we are; and what we choose to
make of ourselves and our world is over to us. We are walking a tight-
rope. Mankind can bring about its own ruin and even destroy this planet.
But there is also within us the potential for a new kind of man. Much of
what men formerly projected into the concept of God is possible of
realisation on earth and from within the human condition. There is a
new sense in which it is possible to say — God comes down on earth.[31]

Reference Notes

1 IN THE BEGINNING WAS THE BOMB

[1] Lloyd Geering, *Faith's New Age*, Collins, 1980
[2] Stephane Groueff, *Manhattan Project*, Little, Brown and Co., 1967
[3] Roger Rapoport, *The Great Atom Bomb Machine*, Dutton, 1971
[4] Robert and Leona Train Rienow, *Our Life With the Atom*, Thomas Y. Crowell and Co., 1959
[5] Document submitted to Dr Kurt Waldheim, September 1980
[6] Michael Howard, "Surviving a Protest", *Encounter*, November 1980
[7] Pierre Wack, "What Makes Some People So Hostile to Large-Scale Projects?" *Shell World*, December 1979
[8] *New Society*, 8 November 1979
[9] *Resurgence*, July/August 1980
[10] Royal Commission on Environmental Pollution, chaired by Sir Brian (now Lord) Flowers, "Report on Nuclear Power and the Environment", HMSO, 1976
[11] *A Blueprint for Survival*, Penguin Special, 1972

2 MURPHY'S LAW AT THREE MILE ISLAND

[1] At Conference on Low-level Ionizing Radiation at Guy's Hospital, London (27 October 1979), organised by the Unit for Study of Health Policy
[2] *Rolling Stone*, 17 May 1979
[3] *Newsweek*, 9 April 1979
[4] Ralph Nader and John Abbotts, *The Menace of Atomic Energy*, Norton and Co., 1977
[5] *Irish Times*, 12 March 1981
[6] *Nuclear Safety Research in the OECD Area: The Response to the Three Mile Island Accident*: OECD Headquarters, Paris, February 1981

3 ECONUTS AND LOOSE SCREWS

[1] *Blackwood's Magazine*, January 1980
[2] ibid.
[3] *Irish Times*, 25 January 1980
[4] ibid., 26 January 1980
[5] ibid.

[6] Nuclear Regulation Reporter, 30, 358/837
[7] Internal NRC Memo., obtained through FOIA by Abalone Alliance
[8] Information obtained through FOIA by the Centre for Law in the Public Interest
[9] ibid.

4 FURTHER ASPECTS OF HUMAN FRAILTY

[1] *Blackwood's Magazine*, January 1980
[2] "The Report of the President's Commission on the Accident at Three Mile Island" (The Kemeny Report) obtainable from the Superintendent of Documents, US Government Printing Office, Washington D C
[3] Robert D. Pollard, ed., "The Nugget File", UCS, Cambridge, Mass.
[4] Walter Patterson, *The Fissile Society*, Earth Resources Research Ltd, 1977
[5] *Rolling Stone*, 17 May 1979
[6] Walter Patterson, op. cit.
[7] *Blackwood's Magazine*, January 1980
[8] *The Village Voice*, 7 May 1979
[9] Walter Patterson, *Nuclear Power*, Pelican, 1976
[10] ibid.
[11] House of Commons Debate, 4 May 1976
[12] House of Lords Debates, 18 and 27 May 1976
[13] *Hampstead and Highgate Express*, 14 March 1980
[14] "Carrying the Can", Report of the Working Party on the Transportation of Nuclear Spent Fuel Through London, London Region Ecology Party, 1980
[15] *Hansard*, 8 November 1979
[16] "Carrying the Can"
[17] Malcolm Stuart, "The Boilermakers who Caused a Nuclear Rethink", *Guardian*, 3 July 1979
[18] "Carrying the Can"
[19] ibid.
[20] ibid.
[21] *Railway Accidents*, Department of Transport
[22] *Irish Times*, 31 July 1979
[23] L.E.J. Roberts, *Radioactive Waste*, UKAEA, 1979
[24] Robert E. Blackith, *The Power that Corrupts*, Dublin University Press, 1976
[25] *Le Monde*, 15 May 1976
[26] *Alternatives au nucléaire*, Press Universitaires, Grenoble, 1975
[27] ibid.
[28] "Energie, croissance", Supplément à UD78, Syndicat des Etudes et Recherches, EDF, 1975
[29] *Irish Times*, 16 December 1980

30 Editorial, *Public Relations Journal*, March 1979
31 *Nucleus*, vol. 1, no.4, May 1979
32 ibid.
33 ibid.
34 *Blackwood's Magazine*, January 1980
35 Helen Caldicott, *Nuclear Madness*, Autumn Press, 1978
36 *Blackwood's Magazine*, January 1980
37 John J. Berger, *Nuclear Power: The Unviable Option*, Ramparts Press, 1976
38 Walter Patterson, *Nuclear Power*, Pelican, 1976
39 *Selected Aspects of Nuclear Power Plant Reliability and Economics*, US General Accounting Office, Washington DC, 15 August 1975
40 Steven Harwood, Kenneth May, Marvin Resnikoff, Barbara Schlenger, Pam Tames, "Decommissioning Nuclear Reactors", NYPIRG, Buffalo, 21 January 1976
41 *A Time to Choose*, Energy Policy Project, Ford Foundation, Cambridge, Mass., 1974
42 W. H. Lunning, "Decommissioning Nuclear Reactors", ATOM 265, November 1979
43 ATOM, September 1979
44 Walter Patterson, *Nuclear Power*, Pelican, 1976
45 Ralph Nader and John Abbotts, *The Menace of Atomic Energy*, Norton and Co., 1977
46 *Nucleus*, vol. II, no. 2, November/December 1979
47 ibid.
48. Ralph Nader and John Abbotts, op. cit.
49 *New York Times*, 10 February 1977
50 Anna Gyorgy, *No Nukes*, South End Press, Boston, Mass., 1979
51 ibid.
52 *New York Times Magazine*, 10 April 1977
53 Anna Gyorgy, op. cit.
54 ibid.
55 Amory Lovins, L. Hunter Lovins, Leonard Ross, *Foreign Affairs*, Summer 1980
56 *Blackwood's Magazine*, January 1980
57 ibid.
58 ibid.
59 Address given at Seabrook, New Hampshire, Rally, 25 June 1978
60 *Blackwood's Magazine*, January 1980
61 *Nuclear Journal Reports*, 22 March 1975

5 HOW IRRADIATED CAN YOU GET?

1 Ralph Nader and John Abbotts, *The Menace of Atomic Energy*, Norton and Co., 1977

² Berger, ed., "Tragedy from 1950 Utah Nuke Tests", *Militant Special Supplement*, 10 April 1979

³ ibid.

⁴ *Observer*, 10 August 1980

⁵ L. J. leVann, "Congenital Abnormalities in Children Born in Alberta During 1961", *Canadian Medical Association Journal*, 89: 120, 1963

⁶ Ralph Nader and John Abbotts, op.cit.

⁷ Peter Bunyard and Gerard Morgan-Grenville; *Nuclear Power: What it Means to you*, Ecoropa, 1980

⁸ Robert Jungk, *The Nuclear State*, John Calder, 1979

⁹ John F. Carroll and Petra Kelly, eds., *A Nuclear Ireland?*, Brindley Dollard, 1978

¹⁰ John Francis and Paul Abrecht, eds., *Facing Up to Nuclear Power*, St Andrew Press, 1976

¹¹ *Irish Times*, 10 January 1981

¹² ibid. 4 March 1981

¹³ A. Stewart, J. Webb, D. Hewitt, "A Survey of Childhood Malignancies", *British Medical Journal*, 1: 1495, 1958

¹⁴ A. Stewart, G.W. Kneale, "Radiation Dose Effects in Relation to Obstetric X-rays and Childhood Cancers", *The Lancet*, 1: 1185, 1970

¹⁵ "Strontium-90 Levels in the Milk and Diet Near Connecticut Nuclear Power Plants", draft report by Ernest J. Sternglass to Congressman D. J. Dodd of Connecticut, 27 October 1977. Quoted in *No Nukes*, Anna Gyorgy, South End Press. Table of Cancer Mortality Rates in Connecticut and New England Before and After Start-up of Millstone Nuclear Plant in Waterford, Connecticut. (Connecticut Health Department, Registration Reports; US Monthly Vital Statistic Reports.)

¹⁶ *Bulletin of Atomic Scientists*, September 1972

¹⁷ John J. Berger, *Nuclear Power: The Unviable Option*, Ramparts Press, 1976

¹⁸ *The Toxicity of Plutonium*, HMSO, 1975

¹⁹ *Health Physics*, vol. 33, no. 5

²⁰ *Washington Post*, 10 February 1977

²¹ *Boston Globe*, 19 February 1978

²² Address given at Rocky Flats, Colorado, April 1978

²³ "Conferences, Conflicts and Cancers", *Vole*, Dec./Jan. 1979/80

²⁴ Hugh Montefiore and David Gosling, eds., *Nuclear Crisis*, Prism Press, 1977

²⁵ Peter Bunyard and Gerard Morgan-Grenville, op. cit.

²⁶ Zhores A. Medvedev, *Nuclear Disaster in the Urals*, Angus & Robertson, 1979

²⁷ ibid.

²⁸ Steve Fetter and Kosta Tsipis, "Catastrophic Nuclear Radiation Releases", Programme in Science and Technology for International Security: Report no. 5, MIT, September 1980

²⁹ ibid.

6 WHAT IS MUF AND WHERE IS IT?

1 *Nuclear Journal Reports*, 22 March 1975
2 John F. Carroll and Petra Kelly, eds., *A Nuclear Ireland?*, Brindley Dollard, Dublin, 1978
3 "US Proliferation Policy Shaken by Simple ORNL Processing Design", *Nucleonics Week*, 27 October 1979
4 Brian Johnson, *Whose Power to Choose?*, International Institute for Environment and Development, 1977
5 ibid.
6 "How Israel Got the Bomb", *Rolling Stone*, 12 January 1977
7 *Sunday Telegraph Magazine*, 1 June 1980
8 ibid.
9 ibid.
10 "Plutonium, Proliferation and Policy", speech given at MIT, 1 November 1976
11 Congressional Records, *Senate*, vol. 121, no. 59, 30 April 1974
12 Mason Willrich and Theodore B. Taylor, *Nuclear Theft: Risks and Safeguards*, Ballinger Books, New York, 1974
13 David Burnham in *New York Times*, 5 August 1977
14 Laurence Martin, ed., *Strategic Thought in the Nuclear Age*, Heinemann, 1979
15 Ralph Nader and John Abbotts, *The Menace of Atomic Energy*, Norton and Co., 1977
16 Charles Komanoff, "Cost Escalation at Nuclear and Coal Power Plants", submitted to *Science*, February 1980; available from Komanoff Energy Associates, New York
17 Amory Lovins, L. Hunter Lovins, Leonard Ross, "Nuclear Power and Nuclear Bombs", *Foreign Affairs*, Summer 1980
18 Fox Butterfield, "Marcos, Facing Criticism, May End $1 billion Westinghouse Contract", *New York Times*, 14 January 1978; Barry Kramer, "Marcos Sets Review of Philippines Award for Nuclear Plant Westinghouse Builds", *Wall Street Journal*, 16 January 1978; "Westinghouse Defends Philippines Role: Marcos Orders Takeover of Disini Firms", *Wall Street Journal*, 17 January 1978
19 W. Kenneth Davis (Bechtel Corporation), interoffice memorandum, 27 September 1973
20 Lovins, Lovins and Ross, op. cit.

7 MAD

1 John Cox, *Overkill*, Penguin, 1977
2 David Boulton, ed., *Voices From the Crowd*, Peter Owen, 1964
3 E. P. Thompson and Dan Smith, eds., *Protest and Survive*, Penguin, 1980
4 John Cox, op. cit.

[5] Helen Caldicott, *Nuclear Madness*, Autumn Press, Brookline, 1978
[6] Howard L. Rosenberg, *Atomic Soldiers*, Beacon Press, Boston, 1980
[7] ibid.
[8] ibid.
[9] ibid.
[10] E. P. Thompson and Dan Smith, op.cit.
[11] John Cox, op. cit.
[12] Paul Erkins and Peter Frings, eds., *How to Survive the Nuclear Age*, The Ecology Party, 1980
[13] David Noble, *The War and Peace Book*, Stockholm International Peace Research Institute, 1977
[14] ibid.
[15] James Garrison, *From Hiroshima to Harrisburg*, SCM Press, 1980
[16] ibid.
[17] Ruth Leger Sivard, *World Military and Social Expenditures*, World Priorities, 1980
[18] ibid.
[19] Anthony Sampson, *The Arms Bazaar*, Coronet, 1977
[20] *Alternative Work for Military Industries*, Richardson Institute for Conflict and Peace Research, 1977
[21] E. P. Thompson and Dan Smith, op. cit.
[22] F. S. L. Lyons, *Culture and Anarchy in Ireland*, Oxford University Press, 1980
[23] Paul Bennett, *Strategic Surveillance*, UCS, Cambridge Mass.
[24] John Cox, op. cit
[25] Dwight D. Eisenhower, *Waging Peace*, Doubleday, 1965
[26] John Cox, op. cit.
[27] Bernard Knight, *Discovering the Human Body*, Heinemann, 1980
[28] Helen Caldicott, op. cit.
[29] ibid.
[30] E. P. Thompson and Dan Smith, op. cit.

8 YESTERDAY'S MEN

[1] Foreword to *Brave New World*, 1946
[2] *Blackwood's Magazine*, October 1979
[3] ibid.
[4] Aldous Huxley, op. cit.
[5] Brian Johnson, *Whose Power to Choose?*, Institute for Environment and Development, 1977
[6] "Contracts for the Supply of Hell-Fire", *Journal of the Law Society of Scotland*, vol. 24, no. 11, November 1979
[7] *Times Service*, 3 April 1981
[8] *Blackwood's Magazine*, October 1979
[9] Federal Radiation Council, Report No. 8, US Government Printing Office, Washington DC

[10] *The Nuclear Fuel Cycle* (Revised Edition, 1975), MIT Press, Cambridge, Mass. and London

[11] Tom Barry, "Uranium Boom in Grants", SEERS, 13–20 December 1975

[12] ibid.

[13] *The Nuclear Fuel Cycle*

[14] Peter Bunyard, "Nuclear Power: The Grand Illusion", *Ecologist*, April/May 1980

[15] ibid.

[16] ibid.

[17] Sir John Hill, "The Reality of Nuclear Power: A Different View", *Blackwood's Magazine*, February 1980

[18] Alun Roberts, *The Rossing File*, CANUC, 1979

[19] ibid.

[20] "The Fast Reactor": leaflet published by Information Services Branch, UKAEA, November 1979

[21] ibid.

[22] Submission at Public Hearings on Projected Commercial Fast Breeder Reactor, London, 13/14 December 1976

[23] ibid.

[24] Jerry J. Cohen, *A Systems Analysis Approach to Nuclear Waste Management Problems*, IIASA Research Memorandum, RM-75-20, May 1975

[25] *Le dossier électronucleaire*, CFDT (Second Edition), 1980

[26] *Nuclear Reactor Safety*, reprinted hearings of JCAE, US Congress, Part 2 vol. I, Washington DC, 1974

[27] David Henderson, "Two Costly British Errors', *Listener*, 27 October 1977

[28] *Irish Times*, 19 February 1981

[29] ibid.

[30] ibid.

[31] Lloyd Geering, *Faith's New Age*, Collins, 1980

Glossary

ABCD: agencies, bureaux, committees and departments (author's acronym).

ACDA: Arms Control and Disarmament Agency (US)

AEA: Atomic Energy Authority (also known as UKAEA).

AEC: Atomic Energy Commission (US); from 1946 to January 1975, the American Federal Government agency with responsibility for the promotion and regulation of nuclear power. Then replaced by the NRC and ERDA.

AGR: advanced gas-cooled reactor; the second-generation design of British reactors.

AIA: American Institute of Architects

ASLAB: Atomic Safety and Licensing Appeals Board (US). Members appointed by the NRC.

BEIR: (Advisory Committee on the) Biological Effects of Ionizing Radiation

BNFL: British Nuclear Fuels Ltd; since 1971 operators of the Windscale reprocessing plant.

B&W: Babcock and Wilcox; designers and suppliers of TMI-2 reactor.

BWR: boiling water reactor

CANDU: Canadian Deuterium-moderated natural-Uranium fuelled reactor

CEGB: Central Electricity Generating Board (UK)

CFR: commercial-scale fast reactor

CIA: Central Intelligence Agency (US)

CLPI: Centre for Law in the Public Interest (US)

Con Ed: Consolidated Edison Co.

DOE: Department of the Environment (UK and US)

EAF: Environmental Action Foundation (US)

ECCS: Emergency Core Cooling System

EEIA: Edison Electric Institute of America

EPA: Environmental Protection Agency (US)

ERDA: Energy Research and Development Administration (US)

Eximbank: Export-Import Bank; a US Federal government agency with responsibility for promoting US exports, including nuclear technology, via loans, loan guarantees and insurance to recipient countries.

FBR: fast breeder reactor

FOE: Friends of the Earth

FOIA: Freedom of Information Act

FRC: Federal Radiation Council (US)

GAO: General Accounting Office (US)

GPU: General Public Utilities Corporation (US)

HARVEST: highly active residues vitrification engineering studies (at Harwell, UK)

HCDA: hypothetical core disruptive accident

HEW: (Department of) Health, Education and Welfare (US)

HSE: Health and Safety Executive (UK)

HTGR: high-temperature gas-cooled reactor

JCAE: Joint Committee on Atomic Energy (US)

IAEA: International Atomic Energy Agency; autonomous, inter-governmental organisation, linked with UN, which promotes and regulates nuclear energy on an international basis.

ICBM: intercontinental ballistic missile

ICRP: International Commission on Radiological Protection; non-governmental organisation founded by scientists to make recommendations on radiation exposure standards.

ID: inventory difference; term now used to describe missing SNM.

IEEE: Institute of Electric and Electronic Engineering (US)

INMM: Institute of Nuclear Materials Management (US)

INPO: Institute for Nuclear Power Operations (US)

IMP: industrial/military/political monolith supporting the development of nuclear weapons and nuclear power; dominant at present in both Communist and Capitalist blocs (author's acronym).

LLL: Lawrence Livermore Laboratory

LMFBR: liquid metal-cooled fast breeder reactor

LOCA: loss-of-coolant accident

LRCM: long range cruise missile

LRL: Lawrence Radiation Laboratory

LWR: light water reactor

MCA: maximum credible accident

Met Ed: Metropolitan Edison Co.; owners of TMI-2.

MIT: Massachusetts Institute of Technology

MUF: material unaccounted for (now replaced by ID).

NASA: National Aeronautics and Space Administration (US)

NCRP: National Council on Radiation Protection (US)

NFS: Nuclear Fuel Services; a subsidiary of Getty Oil (US).

NII: Nuclear Installations Inspectorate (UK)

NPG: Nuclear Power Co. Ltd (UK)

NPT: (Nuclear) Non-Proliferation Treaty

NRC: Nuclear Regulatory Commission (US)

NRPB: National Radiological Protection Board (UK)

NUMEC: Nuclear Materials and Equipment Corporation; now a subsidiary of B&W.

NYPIRG: New York Public Interest Research Group

OECD: Organization for Economic Cooperation and Development

OSA: Official Secrets Act

PFR: prototype fast reactor (the 250 MW reactor at Dounreay).

PG&E: Pacific Gas and Electric Company

PORV: pilot-operated relief valve

PWR: pressurized water reactor

r&d: research and development

RTZ: Rio-Tinto Zinc

SAC: Strategic Air Command

SALT: Strategic Arms Limitation Talks

SAR: Safety Analysis Report; on safety of proposed new nuke required from applicant by NRC before a licence will be issued.

SGHWR: steam generating heavy water reactor

SNM: special nuclear material; uranium enriched in the isotopes 233 or 235; or plutonium. Generally refers to weapons material but can refer to reactor fuel.

SRD: Safety and Reliability Directorate (UKAEA)

SRI: safety-related incidents

SSEB: South of Scotland Electricity Board

TMI: Three Mile Island

TVA: Tennessee Valley Authority; federally owned electric utility serving the Southeastern US; pioneer in the development of nuclear power.

UCS: Union of Concerned Scientists; non-profit-making coalition of scientists, engineers and other professionals concerned about the impact of advanced technology, including nuclear power, on society. Based on Cambridge, Mass., USA.

UNSCEAR: United Nations Scientific Committee on the Effects of Atomic Radiation

WASH-740: 1957 study (The Brookhaven Study) done for the AEC on "Theoretical Possibilities and Consequences of Major Accidents in Nuclear Power Plants". A 1964 revision, which was suppressed, took heed of the larger reactors then being made and estimated 27,000 deaths, $17 billion in damage and contamination of an area the size of Pennsylvania.

WASH-1400: 1974 AEC reactor safety report (The Rasmussen Report) which took three years to complete and cost $4 million. Its reassurances about reactor safety have been widely challenged.

TECHNICAL TERMS

Absorption: process whereby radiation is stopped and reduced in intensity as it passes through matter. The denser the material the better it absorbs, hence the wide use of lead as shielding from radiations (such as X-rays and gamma rays).

Actinides: heavy elements that result from the disintegration of radioactive materials. Actinides produced by fission of reactor fuel are: actinium, thorium, protactinium, uranium, neptunium, plutonium, americium, curium and several more. These must be stored as radioactive wastes.

Afterheat: heat produced from the continual decay of radioactive materials.

Alpha Particle: a positively charged particle of two neutrons and two protons, i.e. the nucleus of a helium atom, which is emitted by certain radioactive material often accompanied by gamma radiation. A notable alpha emitter is plutonium-239, a dangerous carcinogen when ingested or inhaled.

Atomic Wastes: radioactive solids, gases and contaminated liquids produced by the splitting of reactor fuels. Generally classed as high, intermediate or low-level waste depending on curie per litre count.

Beta Particle: high energy electron emitted by decay in a radioactive nucleus. Can cause skin burns and, when ingested, cancer.

Biological Dose: dose of radiation absorbed in biological matter, measured in rems.

Blanket: fuel elements around the core in a fast reactor, containing depleted uranium a part of which is converted to the fissile plutonium-239 by the neutrons escaping from the core.

Bone-seeker: radioactive isotope that tends to accumulate in the bones, such as strontium-90.

Boron: chemical element; powerful absorber of neutrons sometimes used in reactor control rods.

Breed: to form fissile nuclei, usually as a result of neutron capture, possibly followed by radioactive decay. e.g. uranium-238 can capture neutrons that have been slowed (usually by a graphite moderator) to produce neptunium which then decays to plutonium-239, which, unlike uranium-238 is fissile and can be used as an energy source.

Carcinogen: a cancer-causing substance.

Cave: working-space for the manipulation of highly radioactive items; it is surrounded by great thicknesses of concrete or other shielding and has thick protective windows.

Cesium-137: biologically hazardous beta-emitting fission product. It concentrates in muscle tissue and has a half-life of thirty years.

Cladding: material used to cover nuclear fuel (uranium) in order to protect it and to contain the fission products formed during irradiation.

Containment Vessel: large concrete and steel shell around a reactor, designed to contain any radioactivity that might escape from the reactor itself.

Control Rod: rod of a material that absorbs neutrons, used to control the power of a nuclear reactor; usually raised from the core to start a chain-reaction and fully inserted into the core to halt fissioning.

Coolant: liquid (water, molten sodium) or gas (carbon dioxide, helium, air), circulated through a reactor core to remove heat generated there.

Cooling Pond: deep tank of water at nuclear plant sites into which irradiated fuel or defective fuel elements are placed until ready to ship for reprocessing or disposal.

Cooling Tower: massive structure to cool water in a nuke's closed cycle cooling system.

Core: reactor's "heart", containing the fuel and moderator, where the nuclear chain reaction generates heat.

Critical Mass: smallest amount of a fissionable material that will sustain a nuclear chain reaction.

Criticality: state of a nuclear reactor when it achieves a self-sustaining chain-reaction.

Crud: particulate impurities deposited inside nuclear plant circulating water systems.

Curie: unit of radioactivity in general use, corresponding to 37,000 million nuclear disintegrations per second, which is the amount of activity displayed by one gram of radium-226.

Daughter Product: decay product of a radioactive nucleus.

Decay: gradual disintegration of radioactive material over time, through the emission of radiations (usually one or more of alpha, beta, gamma radiation).

Decay Chain: succession of radioactive elements, each formed by the decay of the previous one.

Decay Heat: heat generated by the radioactive decay of the fission products, which continues even after the chain reaction in a reactor has been stopped.

Decontamination: removal of unwanted radioactivity from equipment, surfaces or people.

Depleted Uranium: uranium whose uranium-235 content is less than the 0.71 per cent that occurs in nature.

Dose: amount of energy absorbed by a unit of mass or an organ or individual from irradiation.

Dose Commitment: future radiation doses inevitably to be received, often because a particular radioisotope has been incorporated in body tissues.

Dosimeter: device worn by a person to measure nuclear employees' daily radiation exposure.

Doubling Dose: radiation dose which will cause a doubling of the occurrence of a particular disease in a given population.

Effluent: liquid discharge from a nuke.

Electron: negatively charged atomic particle, about $\frac{1}{2000}$ the mass of a proton or neutron.

Elements: parts of an assembly of nuclear fuel.

Enrichment: process of increasing the concentration of the uranium-235 isotope in uranium beyond 0.71 per cent in order to make it more suitable for use in a reactor.

Environmental Pathway: route by which a radioisotope in the environment is transferred to man, e.g., by biological concentration in foodstuffs.

Fast Neutron: high-energy neutron resulting from atomic fission and travelling at approximately 20,000 kms per second.

Fast Reactor: reactor in which there is no moderator and in which the nuclear chain reaction is sustained by fast neutrons alone.

Fertile: materials such as U-238 and thorium that can become fissionable through neutron absorption.

Film Badge: film packet used for estimating radiation exposure in nuclear facilities.

Fissile: a nucleus that it is capable of undergoing fission if it is struck by and captures a neutron.

Fission: splitting of a heavy nucleus into two or more parts by a neutron, usually accompanied by a release of energy and emission of further neutrons.

Fission Product: nucleus of intermediate size formed from the breakdown or fission of a heavy nucleus such as that of uranium. Nuclei produced by fission of uranium are usually highly radioactive and emit beta particles.

Fuel: material containing fissile nuclei fabricated into a form that can be used in a reactor core.

Fuel Assembly: unit made up of many long fuel rods to be inserted into or removed from the reactor by remote control.

Fuel Cycle: what happens to reactor fuel from mining to waste storage.

Fuel Fabrication: making uranium oxide into fuel pellets and loading fuel rods.

Fuel Pellets: uranium dioxide or other nuclear fuel, in powdered form, that has been ground and pressed into pellets that fit into assembly rods.

Fuel Reprocessing: process of taking highly radioactive spent fuel and extracting the remaining fissionable material through chemical breakdown.

Fusion: merging of two light nuclei to make a heavier one, usually with a release of energy.

Gamma Ray: high energy, short wavelength, electromagnetic radiation emitted by a nucleus. Has the same nature as very high energy X-rays.

Gas Centrifuge: method of isotope separation in which heavier U-238 nuclei are separated from lighter U-235 nuclei by the centrifuging of uranium hexafluoride gas.

Gas-Cooled Reactor: reactor using gas, such as helium, as the coolant.

Gaseous Diffusion: common but costly method of isotope concentration in which U-235 and U-238 molecules in a hexafluoride gas are diffused through thousands of porous metallic membranes, slowly allowing accumulation of the lighter U-235.

Gaseous Wastes: radioactive gases such as krypton-85, iodine-131 and xenon-131, created by nuclear fission.

Genetic Effects: effects produced by radiation in the offspring of the person irradiated, usually malformations.

Gigawatt: one billion watts.

Glove box: sealed box in which workers wearing heavy gloves handle "hot" materials without exposure.

Graphite: black compacted crystalline carbon used, in a very pure form, as a moderator and reflector in some types of reactor core.

Half-Life: time in which the number of nuclei of a particular type is reduced by radioactive decay to one-half. Half-lives can range from minute fractions of a second to many millions of years.

Heavy Water: deuterium oxide. Water containing more than the natural proportion of heavy hydrogen (deuterium) atoms instead of ordinary hydrogen atoms.

Helium: light chemically inert gas used as a coolant in high-temperature gas-cooled reactors.

High-Level: of radioactive waste that requires continuous active cooling to dissipate the internally generated heat and prevent dissemination of the material.

Hydrogen Bomb: nuclear bomb that produces its destructive power largely from atomic fusion; a process triggered by a fission bomb.

Iodine-131: one of the more dangerous fission products, an emitter of beta and gamma radiation with a half-life of 8 days, which tends to accumulate in the thyroid gland.

Ion: atom that has gained or lost one or more electrons and thus become electrically charged.

Isotopes: nuclei of the same chemical element that contain the same number of protons, but different numbers of neutrons (e.g. all nuclei of uranium contain 92 protons, uranium-238 contains 146 neutrons but uranium-235 contains 143 neutrons).

Latent Period: amount of time that elapses between exposure and the first sign of radiation damage.

Leakage: escape of neutrons from a reactor core.

Leukemia: cancer of the blood distinguished by the overproduction of white blood cells. Latent period approximately 5 years. Children are especially vulnerable to it.

Light Water: ordinary water, a compound of hydrogen and oxygen, used as moderator and coolant in some reactors called LWRs.

Megaton: energy of a nuclear explosion equal to one million tons of TNT.

Megawatt: one million watts.

Meltdown: melting of the fuel in the reactor core due to quick uncontrolled temperature rise which would liquefy fuel cladding and rod assemblies.

Millirem (mr): one-thousandth of a rem.

Mixed Oxide: reactor fuel in which the fissile nuclei of plutonium-239 are mixed with natural or reprocessed uranium in a proportion that would equal, in radioactive potential, enriched uranium. Developed for breeder reactors; now suggested for LWRs.

Moderator: substance used to slow neutrons emitted during nuclear fission. Some nuclei, such as uranium-238, can capture slow neutrons, but not fast ones.

Molecule: group of atoms held together by chemical forces.

Mutation: sudden variation, the offspring differing from its parents in one or more heritable characteristics owing to changes within the chromosomes or genes.

Neutron: uncharged particle found in the nucleus of every atom heavier than hydrogen. A free neutron is unstable. With a half-life of 13 minutes it will decay into a proton, electron and neutrino. Ejected at high energy during fission, neutrons sustain a nuclear reactor's chain reaction.

Plutonium-239: heavy, highly toxic, highly carcinogenic radioactive metallic element. It is fissile and used to make nuclear weapons and as breeder reactor fuel. Made in nuclear reactors by bombarding uranium-238 with neutrons.

Power Reactor: reactor for electricity generation, as distinct from weapons material production or research.

Pressure Vessel: large container of welded steel that houses the core of most types of reactors, plus moderator, reflector, thermal shield and control rods.

Price-Anderson Act: Act of US Congress first passed in 1957, and extended in '65 and '75 for ten-year periods, which limits reactor owners' liability to $100 million in the event of an accident and provides for limited Federal indemnity.

Production Reactor: reactor designed to produce by neutron irradiation large amounts of Pu-239 for weapons making.

Proton: part of all nuclei: elementary particle with a single positive electrical charge.

Rad: radiation absorbed dose: the unit of absorbed radiation, corresponding to 0.01 joules of energy per kg of material.

Radiation Sickness: illness induced by exposure to ionizing radiation, ranging in severity from nausea to death.

Radiation Standards: exposure standards for permissible radiation concentrations, rules for safe handling and transportation, and regulations for control of radiation exposure.

Radium: element with 88 protons in the nucleus. An intensely radioactive metallic alpha-emitter, found in nature because it is a decay product of uranium-238. Radium-226 has a half-life of 1622 years.

Radon: alpha-emitting radioactive gas given off by radium, also present during the mining of uranium.

Radon Daughters: the four radioactive short-lived decay products of radon: polonium-218, lead-214, bismuth-214, polonium-214: all metals.

Reactivity: of fuel, its ability to support a nuclear chain reaction; it is a function of the concentration of fissile atoms and inversely of the quantity of neutron-absorbing material present.

Recycling: re-use of fissionable material in irradiated fuel, recovered through reprocessing.

Rem: Roentgen equivalent man: the unit of effective radiation absorbed by tissue; the product of the dose in rads and a quality factor.

Scram: sudden shut-down of fission reaction, usually by remote control insertion of control rods.

Shielding: material interposed between a source of radioactivity and an operator in order to reduce radiation dose.

Somatic Effects: effects produced by radiation in the body of the person irradiated, often cancers.

Spent Fuel: reactor fuel so depleted that it can no longer sustain a chain reaction.

Strontium 90: a radioactive isotope of strontium which is produced by nuclear fission. One cause of bone cancer, it is produced in approximately 6% of all U-235 fissions. It emits high energy beta rays and has a half-life of 28 years. If released into the atmosphere it produces soluble compounds which dissolve in rain water and can be absorbed by plants such as grass. When the contaminated grass is eaten by cows, strontium 90 is concentrated in the milk produced. The alarming increase in strontium 90 content of milk was the side effect of atmospheric nuclear weapons testing programmes which most probably led to their abandonment.

Tailings: fine grey sand left over from uranium-ore milling; contains radium which generates radon.

Tails: depleted uranium produced at an enrichment plant, typically containing only 0.25 per cent of U-235.

Thermal Pollution: heat discharged into the environment; at present some two-thirds of the heat created in a nuke is so discharged.

Uranium: element 92; a heavy metallic, radioactive element. Naturally occurring forms have very long half-lives. Found in nature as a mixture of the isotopes uranium-235 (0.71 per cent) and uranium-238 (99.3 per cent). U-235 and the artificially produced U-233 are fissile. U-238 is fertile.

Uranium Hexafluoride: gas made from yellowcake and used in the gaseous diffusion enrichment process.

Vitrification: fusing of high-level waste into glass-like solids for long-term storage.

Weapons Grade: of uranium or plutonium, capable of being made into a nuclear assembly that would be critical on fast prompt neutrons alone.

Yellowcake: raw materials of nuclear fuel: mixture of the two oxides of uranium.

Zincaloy: alloy of zirconium used as nuclear fuel cladding because it tends not to absorb neutrons.

Bibliography

BOOKS

Leonard Beaton, *Must the Bomb Spread?*, Pelican Original, 1966

John J. Berger, *Nuclear Power: The Unviable Option*, Ramparts Press, Palo Alto, 1976

David Boulton, ed., *Voices From the Crowd: Against the H-Bomb*, Peter Owen, 1964

Ian Breach, *Windscale Fallout: A Primer for the Age of Nuclear Controversy*, Penguin Special, 1978

Peter Bunyard and Gerard Morgan-Grenville, *Nuclear Power: What it Means to You*, Ecoropa, 1980

Nigel Calder, ed., *Unless Peace Comes: A Scientific Forecast of New Weapons*, Pelican, 1968

Helen Caldicott, *Nuclear Madness*, Autumn Press, Brookline, Mass., 1978

John F. Carroll and Petra K. Kelly, eds., *A Nuclear Ireland?*, Irish Transport and General Workers' Union, 1978

Zdenek Cervenka and Barbara Rogers, *The Nuclear Axis: Secret Collaboration between West Germany and South Africa*, Julian Friedmann Books, London, 1980

Peter Chapman, *Fuel's Paradise: Energy Options for Britain*, Pelican Original, 1980

John Cox, *Overkill: The Story of Modern Weapons*, Peacock Books, 1977

Dave Elliott, with Pat Coyne, Mike George and Roy Lewis, *The Politics of Nuclear Power*, Pluto Press, 1978

William Epstein and Toshiyuki Toyoda, eds., *A New Design for Nuclear Disarmament*, Pugwash Symposium, Kyoto, Japan, The Bertrand Russell Peace Foundation Ltd, 1977

Gerald Foley with Charlotte Nassim; consultant Gerald Leach, *The Energy Question*, Pelican Original, 1976

John Francis and Paul Abrecht, eds., *Facing Up to Nuclear Power: A Contribution to the Debate on the Risks and Potentialities of the Large-Scale Use of Nuclear Energy*, St Andrew Press, Edinburgh, 1976

Anna Gyorgy and friends, *No Nukes: Everyone's Guide to Nuclear Power*, South End Press, Boston, 1979

John Hersey, *Hiroshima*, Penguin Modern Classics, 1980

Fred Hoyle and Geoffrey Hoyle, *Commonsense in Nuclear Energy*, Heinemann, 1980

Ivan D. Illich, *Energy and Equity*, Calder and Boyars, London, 1974

Claire Jones, Steve J. Gadler, Paul H. Engstrom, *Pollution: The Dangerous Atom*, Dent, 1972

Robert Jungk, *The Nuclear State*, John Calder, London, 1979

Laurence Martin, ed., *Strategic Thought in the Nuclear Age*, Heinemann, 1979

Zhores A. Medvedev, *Nuclear Disaster in the Urals*, Angus and Robertson, 1979

Hugh Montefiore and David Gosling, eds., *The Nuclear Crisis: A Question of Breeding*, Prism Press, Dorchester, for the British Council of Churches, 1977

Ralph Nader and John Abbotts, *The Menace of Atomic Energy*, Norton and Co., New York, 1977

David Noble, *The War and Peace Book*, Writers' and Readers' Publishing Cooperative, London, 1977

Walter C. Patterson, *The Fissile Society: Energy, Electricity and the Nuclear Option*, Earth Resources Research Ltd, 1977

Walter C. Patterson, *Nuclear Power*, Pelican Original, 1976

David Pearce, Lynne Edwards, Geoff Beuret, *Decision Making for Energy Futures: A Case Study of the Windscale Inquiry*, Macmillan Press, 1979

Barry Popkess, *The Nuclear Survival Handbook: Living Through and After a Nuclear Attack*, Arrow Books, 1980

Howard L. Rosenberg, *Atomic Soldiers: American Victims of Nuclear Experiments*, Beacon Press, Boston, 1980

Theodore Roszak, *The Making of a Counter Culture*, Faber Paperback, 1970

Anthony Sampson, *The Arms Bazaar*, Hodder and Stoughton, 1977

Nicholas A. Sims, *Approaches to Disarmament: An Introductory Analysis*, Quaker Peace and Service, 1979 (Revised and Expanded Edition)

E. P. Thomson, *Protest and Survive*, Bertrand Russell Peace Foundation, 1980

Barbara Ward, *The Home of Man*, André Deutsch, 1976

Barbara Ward, *Progress for a Small Planet*, Pelican Original, 1979

Roger Williams, *The Nuclear Power Decisions*, Croom Helm, London, 1980

Elizabeth Young, *A Farewell to Arms Control?*, Pelican Original, 1972

From the *Ecologist*, *A Blueprint for Survival*, Penguin Special, 1972

Disarm or Die: A Disarmament Reader for the Leaders and Peoples of the World, Taylor and Francis, London, 1978

PAMPHLETS, REPORTS, ETC.

"Alternative Work for Military Industries", Dave Elliott, Mary Kaldor, Dan Smith, Ron Smith, Richardson Institute for Conflict and Peace Research, London, 1977

"Anti-Nuclear Now or ... Never", Students Against Nuclear Power, 1980

"The Big Risk: Nuclear Power", Michael Flood, FOE, 1980

"Bombs for Breakfast", Sandy Merritt, Tom Reed, Adrian Walker, War on Want, 1978

"Carrying the Can": Report of the Working Party on the Transportation of Nuclear Spent Fuel through London, Ecology Party, North London Branch, 1980

"The Easy Path Energy Plan", Vince Taylor, Union of Concerned Scientists, Cambridge, Mass., 1979

"Energy in Question": Report of a Working Party on the Future Sources of Energy for Ireland, An Taisce, 1979

"Hell NO, We Won't Glow: Nonviolent Occupation of a Nuclear Power Site", Sheryl Crown, Clamshell Alliance, New Hampshire, 1979

"How to Survive the Nuclear Age", Paul Ekins and Peter Frings, the Ecology Party, 1980

"Looking But Not Seeing": The Federal Nuclear Power Plant Inspection Programme, Lawrence S. Tye, Union of Concerned Scientists, Cambridge, Mass., 1978

"No Cruise Missiles: No SS20s", Ken Coates, Bertrand Russell Peace Foundation, 1980

"Non-Nuclear Energy Options for the UK", SERA Energy Group, 1980

"The Nuclear Fuel Cycle: A Survey of the Public Health, Environmental, and National Security Effects of Nuclear Power", MIT Press, Cambridge, Mass., and London, England. Revised Edition, 1975

"Nuclear Prospects: A Comment on the Individual, the State and Nuclear Power", Michael Flood and Robin Grove-White, FOE/C-PRE/NCCL, London, 1976

"Nuclear Reactor Licensing: A Critique of the Computer Safety Prediction Methods", Carl J. Hocevar, Union of Concerned Scientists, Cambridge, Mass., 1975

"The Nuggett File": Excerpts from the Government's Special Internal File on Nuclear Power Plant Accidents and Safety Defects, Robert D. Pollard, ed., Union of Concerned Scientists, Cambridge, Mass., 1979

"The Politics of Ecology", the Ecology Party, London, 1980

"The Power that Corrupts: The Threat of Nuclear Power in Ireland", Robert E. Blackith, Dublin University Press, 1976

"Radiation Protection": Recommendations of the International Commission on Radiological Protection (adopted 17 January 1977). ICRP publication 26, Pergamon Press

"Radiation: Your Health at Risk", Radiation and Health Information Service, Cambridge, 1980

"Red Light for Yellowcake: The Case Against Uranium Mining", Denis Hayes, Jim Falk, Neil Barrett, FOE Australia, 1977

Report on Nuclear Power and the Environment, the Royal Commission on the Environment, chaired by Sir Brian (now Lord) Flowers, HMSO, 1976 [The Flowers Report]

Report on the President's Commission on the Accident at Three Mile Island (The Kemeny Report) obtainable from the Superintendent of Documents, US Printing Office, Washington DC

"The Risks of Nuclear Power Reactors": A Review of the NRC Reactor Safety Study WASH-1400, Union of Concerned Scientists, Cambridge, Mass., 1977

"South Africa's Nuclear Capability", Dan Smith, UN Centre Against Apartheid, 1980

"Strategic Surveillance: How America Checks Soviet Compliance with SALT", Paul Bennett, Union of Concerned Scientists, Cambridge, Mass., 1979

Torness Alliance Occupier's Handbook, Torness Alliance, 1979

"Torness: Keep it Green", Michael Flood, FOE

"Uranium Mining in Donegal: Its Dangers and Deceits", Just Books, Belfast, 1979

"What Choice Windscale?: The Issues of Reprocessing", Czech Conroy, FOE and the Conservation Society, London, 1978

"Whose Power to Choose? International Institutions and the Control of Nuclear Energy", Brian Johnson, International Institute for Environment and Development, London, 1977

"World Military and Social Expenditures: 1980", Ruth Leger Sivard, World Priorities, Box 1003, Leesburg, Virginia, USA, 1980

Index